SpringerBriefs in Applied Sciences and Technology

SpringerBriefs present concise summaries of cutting-edge research and practical applications across a wide spectrum of fields. Featuring compact volumes of 50 to 125 pages, the series covers a range of content from professional to academic.

Typical publications can be:

- A timely report of state-of-the art methods
- An introduction to or a manual for the application of mathematical or computer techniques
- A bridge between new research results, as published in journal articles
- A snapshot of a hot or emerging topic
- An in-depth case study
- A presentation of core concepts that students must understand in order to make independent contributions

SpringerBriefs are characterized by fast, global electronic dissemination, standard publishing contracts, standardized manuscript preparation and formatting guidelines, and expedited production schedules.

On the one hand, **SpringerBriefs in Applied Sciences and Technology** are devoted to the publication of fundamentals and applications within the different classical engineering disciplines as well as in interdisciplinary fields that recently emerged between these areas. On the other hand, as the boundary separating fundamental research and applied technology is more and more dissolving, this series is particularly open to trans-disciplinary topics between fundamental science and engineering.

Indexed by EI-Compendex, SCOPUS and Springerlink.

More information about this series at http://www.springer.com/series/8884

Diego T. Santos · Ádina L. Santana ·
M. Angela A. Meireles · M. Thereza M. S. Gomes ·
Ricardo Abel Del Castillo Torres ·
Juliana Q. Albarelli · Aikaterini Bakatselou ·
Adriano V. Ensinas · François Maréchal

Supercritical Fluid Biorefining

Fundamentals, Applications and Perspectives

 Springer

Diego T. Santos
LASEFI/DEA, School of Food Engineering
University of Campinas—UNICAMP
Campinas, São Paulo, Brazil

M. Angela A. Meireles
LASEFI/DEA, School of Food Engineering
University of Campinas—UNICAMP
Campinas, São Paulo, Brazil

Ricardo Abel Del Castillo Torres
LASEFI/DEA, School of Food Engineering
University of Campinas—UNICAMP
Campinas, São Paulo, Brazil

Aikaterini Bakatselou
IPESE
École Polytechnique fédérale de
Lausanne-EPFL
Sion, Valais, Switzerland

François Maréchal ⓘ
IPESE
École Polytechnique fédérale de
Lausanne-EPFL
Sion, Valais, Switzerland

Ádina L. Santana
LASEFI/DEA, School of Food Engineering
University of Campinas—UNICAMP
Campinas, São Paulo, Brazil

M. Thereza M. S. Gomes
LASEFI/DEA, School of Food Engineering
University of Campinas—UNICAMP
Campinas, São Paulo, Brazil

Juliana Q. Albarelli
LASEFI/DEA, School of Food Engineering
University of Campinas—UNICAMP
Campinas, São Paulo, Brazil

Adriano V. Ensinas ⓘ
IPESE
École Polytechnique fédérale de
Lausanne-EPFL
Sion, Valais, Switzerland

ISSN 2191-530X   ISSN 2191-5318   (electronic)
SpringerBriefs in Applied Sciences and Technology
ISBN 978-3-030-47054-8   ISBN 978-3-030-47055-5   (eBook)
https://doi.org/10.1007/978-3-030-47055-5

This Springer imprint is published by the registered company Springer Nature Switzerland AG
The registered company address is: Gewerbestrasse 11, 6330 Cham, Switzerland

Introduction

An important challenge for the food and agricultural industry is the reduction or extinction of wastes through intensified use of raw material for the obtaining of high value-added products. The employment of clean and economic feasible technologies, such as supercritical fluid-based technologies, contribute to the recovery, and valorization of such materials for inclusion in the human diet, as functional ingredients, and/or for biorefinery purposes, as cheaper resources. Thus, this book aims to provide deep insights into the fundamentals, applications, and perspectives of the use of supercritical CO_2 as solvent and antisolvent for biorefining.

Chapter 1, entitled "Supercritical Fluid Biorefining Using Supercritical CO_2 as an Antisolvent for Micronization, Coprecipitation, and Fractionation: Fundamentals, Processing, and Effect of Process Conditions", provides insights about the fundamentals and effect of operational conditions of supercritical CO_2 antisolvent-based processes, such as Supercritical Antisolvent (SAS) Precipitation and Supercritical Antisolvent Fractionation (SAF) processes.

In Chap. 2, entitled "Supercritical Fluid Biorefining Using Supercritical CO_2 as an Antisolvent for Micronization, Coprecipitation, and Fractionation: Recent Applications", it is then discussed the recent (2013–2018) applications on the use of these supercritical fluid antisolvent-based processes. They are able to microencapsulate and/or purify many substances that are difficult to treat with conventional techniques. In addition, the control of the morphology of materials by adjusting nucleation and growth during particle production is provided. Among the applications, this chapter focuses on the use of supercritical CO_2 as an antisolvent for micronization, coprecipitation, and fractionation of high-value products for the food, cosmetic, and pharmaceutical industries, most focusing on the applications of integrated techniques on the biorefining of plant matrices into marketable products.

Chapters 3 and 4 present some perspectives about the economics and process integration with other processes aiming at the development of novel conceptual biorefining approaches for plant materials valorization. In addition, both chapters present experimental and mathematical simulation results regarding novel approaches and integrated biorefinery concepts using supercritical fluids as an antisolvent and as a solvent, respectively.

Chapter 3, entitled "Integrated Biorefinery Approach for the Valorization of Plant Materials Using Supercritical Antisolvent-Based Precipitation Technique for Obtaining Bioactive Compounds", investigates a novel approach for turmeric rhizomes valorization for the obtaining of microparticles composed of curcuminoids ethanolic extract and fractionated volatile oils, using SAS precipitation process using supercritical carbon dioxide as an antisolvent and the recovery of starch, curcuminoids, and glucose using pressurized hot water. In addition, a cheap and versatile method for the quantification curcuminoids using thin-layer chromatography coupled to image processing analysis was applied to the solid wastes and liquid extracts from turmeric, derived from extraction processes which employed supercritical CO_2 and pressurized liquid ethanol.

Meanwhile, Chap. 4, entitled "Perspectives on Vanillin Production from Sugarcane-Bagasse Lignin Using Supercritical CO_2 as a Solvent in a Novel Integrated Second-Generation Ethanol Biorefinery", focus on the use of supercritical CO_2 as a solvent for the supercritical fluid extraction (SFE) of vanillin from organosolv media. The use of sugarcane bagasse as a biomass source to product diversification has gaining much attention recently and it is a very promising research topic.

Finally, Chap. 5, entitled "Novel Biorefinery Concept for the Production of Carotenoids from Microalgae Using Lignocellulose-Based Biorefinery Products and Supercritical Fluids", proposes an innovative biorefinery conceptual process for the production of carotenoids from microalgae using lignocellulose-based biorefinery products and/or by-products and pressurized fluids. The extraction process, which can be done also with microalgal biomass with high content of moisture avoiding high-cost downstream processes, involves the use of ethanol and 2-MethylTetraHydroFuran (2 MTHF) mixed or in a sequential form for selective extraction of carotenoids. 2 MTHF is obtained from furfural, which is produced as a by-product during lignocellulosic biomass (sugarcane bagasse, wood, corn stover, rice straw, etc.) pre-treatment for ethanol production, for example. The solvent recovery step involves the use of CO_2, which is obtained from ethanol fermentation as a by-product. Specific conditions for CO_2 for temperature and pressure to achieve supercritical conditions would be applied in order to besides high solvent recovery and recycling provide a desirable selective carotenoid purification and encapsulation if a coating material is added. The two-step process can be converted in a one-step process minimizing carotenoid degradation if the extraction process is performed under higher pressure than that performed during extract precipitation. In addition, the proposed processing route can be well integrated into conventional existing biofuels production (gasification, combustion, etc.) scenarios using the solids recovered after carotenoids production as feedstock.

Considering that Brazil, among other countries, have enormous potential for biorefining of renewable raw materials, this book also has the aim of designing feasible economically integrated supercritical fluid biorefineries, in which the SAS, SAF, and SFE process should have an important task to complete.

Campinas, Brazil

Diego T. Santos
diego_tresinari@yahoo.com.br
Ádina L. Santana
adina.santana@gmail.com
M. Angela A. Meireles
maameireles@lasefi.com

Contents

1 **Supercritical Fluid Biorefining Using Supercritical CO$_2$ as an**
  Antisolvent for Micronization, Coprecipitation, and Fractionation:
  Fundamentals, Processing, and Effect of Process Conditions   1
  1.1   Introduction   1
  1.2   Supercritical Antisolvent (SAS) Precipitation Fundamentals   3
  1.3   SAS/SAF Operational Processing   3
  1.4   Effect of Operational Conditions in SAS/SAF Process   5
        1.4.1   Temperature and Pressure   6
        1.4.2   Solute Concentration   7
        1.4.3   Nozzle Geometry.   8
        1.4.4   CO$_2$ and Solution Flow Rates.   8
        1.4.5   Type of Organic Solvent   9
  1.5   Conclusions   10
  References   10

2 **Supercritical Fluid Biorefining Using Supercritical CO$_2$ as an**
  Antisolvent for Micronization, Coprecipitation, and Fractionation:
  Recent Applications   13
  2.1   Introduction   13
  2.2   Applications for SAS   14
  2.3   Applications for SAF   14
  2.4   Conclusions   14
  References   31

3 **Integrated Biorefinery Approach for the Valorization of Plant**
  Materials Using Supercritical Antisolvent-Based Precipitation
  Technique for Obtaining Bioactive Compounds   33
  3.1   Introduction   33
  3.2   Materials and Methods   34
        3.2.1   Materials.   34
        3.2.2   Supercritical Fluid and Pressurized Liquid Extraction.   34

3.2.3   Coprecipitation of Turmeric Extracts   35
3.2.4   Scanning Electron Microscopy (SEM)   35
3.2.5   Thin-Layer Chromatography   35
3.3   Results and Discussion   37
3.3.1   Scanning Electron Microscopy Analysis   37
3.4   Thin-Layer Chromatography Coupled to Image Estimation
Background   37
3.4.1   Method Validation   37
3.4.2   Precision and Accuracy   43
3.4.3   Application of the Method to Turmeric Products   43
3.5   Conclusions   46
References   46

4   **Perspectives on Vanillin Production from Sugarcane Bagasse
Lignin Using Supercritical CO₂ as a Solvent in a Novel Integrated
Second-Generation Ethanol Biorefinery**   49
4.1   Introduction   49
4.2   Materials and Methods   50
4.2.1   Biorefinery of Sugarcane Considered to the Production
of High-Quality Lignin   50
4.2.2   Vanillin Production Process   51
4.3   Results and Discussion   53
4.4   Conclusions   55
References   55

5   **Novel Biorefinery Concept for the Production of Carotenoids
from Microalgae Using Lignocellulose-Based Biorefinery Products
and Supercritical Fluids**   57
5.1   Introduction   57
5.2   Prior Art Searches   58
5.2.1   Integral Use of Algal Biomass   58
5.2.2   Sequential Extraction Using Pressurized Fluids   60
5.2.3   Integration of the Extraction Process to Purification
and/or Encapsulation Processes   64
5.2.4   Integration of New Processes into Existing Industrial
Facilities   67
5.3   Description of the Proposed Supercritical Fluid Biorefinery
Concept   68
5.4   Conclusions   70
References   71

About the Authors

Diego T. Santos holds his Ph.D. in Food Engineering from the University of Campinas (UNICAMP, Brazil) in 2011 and a BS degree in Chemical Engineering from the University of São Paulo (USP, Brazil), 2008. Since 2011, he is working as a Scientific Researcher in the food engineering department at the UNICAMP. Between May 2013 and April 2014, he did post-doctoral internships at the Swiss Federal Institute of Technology (EPFL) with Prof. Dr. François Maréchal and at University of Valladolid (Spain) with Prof. Dr. Maria Jose Cocero. Besides, he did a short-term research period at Dublin City University (Ireland), at the University of Chile (Chile), and at CONICET-Bahía Blanca (Argentina). He has published 85 papers in peer-reviewed journals, 15 book chapters, and more than 125 works in scientific conferences. Moreover, he developed 11 new processes, has 1 patent, and 10 scientific awards/nominations. He has participated in more than 45 research projects with both, public and private funding. He has supervised two Ph.D. theses, six M.Sc. dissertations, and ten undergraduate research projects. His research is related to biomass valorization through the use of clean technologies. Among his many activities related to the promotion of scientific development, he serves as Reviewer for 70 international journals and as member of the editorial advisory board for 17. In addition, he also was part of the organizing committee of 7 scientific conferences. He was Lead Guest Editor for a special edition of the *International Journal of Chemical Engineering* (Hindawi) and has Edited the book *Supercritical Antisolvent Precipitation Process: Fundamentals, Applications, and Perspectives* (Springer Nature). He keeps scientific collaboration with several institutions: The Energy and Research Institute (TERI, Northeast Regional Centre, India), Universidad de Carabobo (Venezuela), Universidad Técnica de Machala (Ecuador), University of Valladolid (Spain), Federal University of Rio Grande do Norte (UFRN, Brazil), State University of Feira de Santana (Brazil), Federal Institute of Education, Science and Technology (Capivari Campos, Brazil), among others.

Ádina L. Santana holds a Ph.D. in Food Engineering (2017) from the University of Campinas (UNICAMP, Brazil), an M.Sc. in Chemical Engineering (2012), and a BS Degree in Food Engineering (2011) from Federal University of Pará (UFPA, Brazil). She worked during 2018–2019 as a Postdoctoral Researcher Associate in the Food and Nutrition Department at the UNICAMP (Brazil) in the bioprocesses expertise coupled with the use of clean technologies to obtain products with enhanced quality for human health and nowadays she is working as a Postdoctoral Researcher Associate at the University of Nebraska (USA). She serves also as a Reviewer for 29 international journals. In addition, she also served as a Reviewer for the 5th International Conference on Agricultural and Biological Sciences—ABS 2019 (Macau, China). She has published 31 research papers in peer-reviewed journals, 16 book chapters, and 17 works in scientific conferences. In addition, she has edited the book *Supercritical Antisolvent Precipitation Process: Fundamentals, Applications, and Perspectives* (Springer Nature). She has knowledge in extraction and encapsulation of bioactive compounds with the use of supercritical fluids, pressurized liquids, and enzymes.

M. Angela A. Meireles is Director of Innovation of Natural Bioactive, Coordinator of Food Science of Coordination of Superior Level Staff Improvement (CAPES), and Professor from the Food Engineering Department at the UNICAMP (Brazil), where she began working in 1983 as Assistant Professor. In 2016, she retired from the UNICAMP where still holds the position of Professor for the Post-Graduate Program in Food Engineering and is the leading research of three technology transfer projects: (1) development of a process to obtain an extract from Cannabis sativa, for Entourage Lab (http://entouragelab.com/); (2) development of an integrated process to produce bioactive for cosmetic industry (https://pt-br. facebook.com/scosmeticosdobem/); and (3) assembling a supercritical fluid pilot plant to process Algae, for Cietec (http://www.cietec.org.br/project/bioativos/). During her time in Academia, she taught among other course thermodynamics, mass transfer, design, and so on at the undergraduate and graduate level. She worked closely with industries developing generally recognized as safe (GRAS) processes to obtain extracts from a variety of natural resources. After her retirement from the UNICAMP, she became a Business Partner of Bioativos Naturais. She holds a Ph.D. in Chemical Engineering from Iowa State University (USA, 1982), and an M.Sc. (1979) and BS degrees (1977) in Food Engineering both from UNICAMP. She has published over 348 research papers in peer-reviewed journals and has made more than 500 presentations at scientific conferences. She has supervised 47 Ph.D. theses, 31 M.Sc. theses, and approximately 68 undergraduate research projects. Her research is in the field of the production of extracts from aromatic, medicinal, and spice plants by supercritical fluid extraction and conventional techniques, such as steam distillation and GRAS solvent extraction, including the determination of process parameters, process integration, and optimization, extracts' fractionation, and techno-economical analysis of the process. She has coordinated scientific exchange projects between UNICAMP and European universities in France, Germany, Holland, and Spain. Nationally, she coordinated a project (SuperNat) that involved 6 Brazilian institutions (UNICAMP, UFPA, UFRN, UEM, UFSC, IAC) and a German university (Technishe Universität Hamburg-Harburg—TUHH). In 2000–2005, she coordinated a thematic project financed by FAPESP (State of São Paulo Science Foundation) on supercritical technology applied to the processing of essential oils, vegetable oils, pigments, stevia, and other natural products. She has coordinated 4 technology transfer projects in supercritical fluid extraction from native Brazilian plants. She coordinated two projects in supercritical fluid chromatography to analyze petroleum in a partnership with Petrobras and to analyze the food system in a partnership with the Waters Technologies of Brazil. She is Editor-in-Chief of The Open Food Science Journal (https://benthamopen.com/TOFSJ/home/). She is also Associate Editor of Food Science and Technology—Campinas (www.scielo.br/cta), RSC Advances (https://www.rsc.org/journals-books-databases/about-journals/rsc-advances/) and Heliyon Food Science (https://www.sciencedirect.com/journal/heliyon). She belongs to the editorial boards of the Journal of Supercritical Fluids, Journal of Food Processing Engineering (Blackwell Publications), Recent Patents on Engineering (Bentham Science Publications), The Open Chemical Engineering Journal (Bentham Science Publications), Pharmacognosy Reviews (Pharmacognosy Networld), Food and Bioprocess Technologies (Springer). She was the COEIC of Recent Patents on Engineering from 2016 to 2017. From 1994 to 1998, she served as Associate Editor for the journals Food Science and Technology (Campinas) and Boletim do SBCTA (Newsletter from the Brazilian Society of Food Science and Technology). She is the editor of three books: (1) *Extracting Bioactive Compounds for Food Products: Theory and Application* (CRC Press, Boca Raton, USA), (2) *Fundamentos de Engenharia de Alimentos* (Food Engineering Fundamentals—Atheneu, São Paulo, Co-editor Dr. C. G Pereira), and (3) *Supercritical Antisolvent Precipitation Process: Fundamentals, Applications, and Perspectives* (Springer Nature, Cham, Switzerland). She was a Guest Editor of special issues of the *Journal of Supercritical Fluids* and *The Open Chemical Engineering Journal*.

M. Thereza M. S. Gomes is a Professor of thermal and fluids at Mackenzie Presbyterian University. She holds a Ph.D. in Food Engineering from UNICAMP, in the area of Physical Separations, funded by National Council for Scientific and Technological Development (CNPq), with a sandwich doctorate from the University of Alberta, funded by the PDSE-CAPES program. She holds a master's degree in Food Engineering from UNICAMP, in the area of process engineering applied to the food industry, funded by the Coordination of Superior Level Staff Improvement (CAPES). She holds a degree in Food Engineering from the University of Taubaté (2008) and has completed three scientific initiation projects funded by FAPESP.

Ricardo Abel Del Castillo Torres is a Professor in the Food Engineering Department (DIA)/Faculty of Food Industries (FIA) from the National University of the Peruvian Amazon (Iquitos, Peru). He holds a Ph.D. in Food Engineering from the University of Campinas (UNICAMP, Brazil) in 2019 with the financial support from the Coordination of Superior Level Staff Improvement (CAPES, Financial code 001). In 2015, he received his M.Sc. degree in Food Engineering at the UNICAMP with financial support from the National Council for Scientific and Technological Development (CNPq). He was the winner of the Leopold Hartman Award, attributed from his work presented at the XXVI Brazilian Congress of Food Science and Technology (XXVI CBCTA), in the oils and fats category. He has experience in the area of Food Science and Technology, with emphasis on Food Engineering, acting on themes of environmental impact reduction through the recycling of pressurized CO_2 in pilot scale, the obtaining diversified products from pressurized fluid processes, optimization and process integration, extraction with supercritical fluids, precipitation with supercritical antisolvent, application of ultrasound technology to obtain bioactive compounds, and processes for equipment design and industrial systems.

Juliana Q. Albarelli holds a BS degree in Chemical Engineering from the University of São Paulo (USP) (2003–2008), an M.Sc. degree in Chemical Engineering (2008–2009), and a Ph.D. in Chemical Engineering from the University of Campinas (UNICAMP, 2009–2013) with a sandwich internship at the Universidad de Valladolid (Spain) and Post-doctorate at the École Polytechnique Fédérale de Lausanne (Switzerland) (2013–2014) and at the Universidad de Valladolid (Spain, 2016). Since 2006, she has been developing research, development, and innovation activities. She acts as a Reviewer for nine international journals. She has published 44 articles in specialized journals, 8 book chapters, and 46 papers in national and international scientific events. In addition, she has experience in using computational tools for evaluation and optimization of emerging processes. Additionally, she has experience in University–Company interaction projects. She participated in 34 events in Brazil and abroad, participating in the organization of 2 of them. She is currently working as a Post-doctoral Researcher at the School of Food Engineering at UNICAMP. She maintains technical and scientific collaboration with several national and foreign universities such as University of São Paulo (Campus Lorena, Brazil), University of Valladolid (Spain), École Polytechnique Fédérale de Lausanne (Swiss), and the Federal Institute of Education, Science and Technology (Capivari Campos, Brazil).

Aikaterini Bakatselou holds a Bachelor and Master of Science in Chemical Engineering from Aristoteleion Panepistimion Thessalonikis (Greece) (2006–2012) and Masters degree, Management of Technology and Entrepreneurship from École Polytechnique fédérale de Lausanne (EPFL, Switzerland) (2012–2014). Current she is working as Crude Oil Trading Operator at Total Company (Switzerland).

Adriano V. Ensinas graduated at Mechanical Engineering from University of Campinas (2001), got a master's at Mechanical Engineering from University of Campinas (2003), and Ph.D. at Mechanical Engineering from University of

Campinas (2008). He has experience in Mechanical Engineering, focusing on Thermal Engineering, acting on the following subjects: modeling and process integration, cogeneration systems, biomass, energy conversion, and biofuels production. He has published 52 papers in peer-reviewed journals, 6 book chapters and more than 45 works in scientific conferences. Currently, he is working as an adjunct professor at the Federal University of Lavras (Brazil). He serves as a Reviewer for several international journals and as Associate Editor for *Frontiers in Energy Research*. Since 2012, he is the associate researcher to École Polytechnique Fédérale de Lausanne (Swtizerland).

François Maréchal holds a process engineering degree (1986) and a Ph.D. from the University of Liège in Belgium (1995), where he realized a Ph D. in the field of process integration of industrial sites under the supervision of Prof. B. Kalitventzeff. In 2001, he moved to École Polytechnique Fédérale de Lausanne (EPFL) in Switzerland where he joined the Industrial Energy Systems Laboratory of Prof. D. Favrat. Since 2013, he is a professor in EPFL in the EPFL Valais-Wallis Campus, heading the Industrial Process and Energy Systems Engineering group. He is conducting research in the field of the process and energy systems engineering for the rational use of energy and resources in industrial processes and energy systems. His activities are focussing on the development of computer-aided process and energy system design methods applying process integration and optimization techniques. He has produced more than 394 scientific papers (with H-index of 40) in the field of energy efficiency in the industry, process system design for biofuels and electricity production, industrial ecology and sustainable energy systems in urban areas studying the optimal integration of renewable energy resources for the energy transition. He has been a member of the scientific committee of IFP Energies Nouvelles (F) and is now an expert to the scientific committee of IFP Energies Nouvelles. François Maréchal is co-chair of the energy section of the European Federation of Chemical Engineering and the representative of Switzerland in the Working Party on the Use of Computers in Chemical Engineering of the European Federation of Chemical Engineering. François Maréchal is specialty chief editor of Frontiers in Energy Research, Specialty Process, and Energy Systems Engineering. In EPFL, François Maréchal is teaching at bachelor, master and doctoral school levels in mechanical engineering and the minor in Energy a multidisciplinary program in Energy offered to the master of EPFL. He has directed or co-directed 44 Ph.D. thesis and directed more than 200 Master thesis. Publications of François Maréchal may be found on http://www.researcherid.com/rid/B-5685-2009.

Chapter 1
Supercritical Fluid Biorefining Using Supercritical CO$_2$ as an Antisolvent for Micronization, Coprecipitation, and Fractionation: Fundamentals, Processing, and Effect of Process Conditions

Abstract The use of supercritical CO$_2$ (SC–CO$_2$) antisolvent for micronization, coprecipitation, and fractionation of high-value products for biorefining of plant matrices into marketable products has been a promising and increasing research topic. These SC–CO$_2$ antisolvent processes are able to microencapsulate many materials that are difficult to treat with conventional techniques. In addition, the control of the morphology of materials by adjusting nucleation and growth during particle production is provided. The use of supercritical antisolvent processes is advantageous when compared with other methods like freeze-drying, drying at high temperatures and spray-drying such as uniform particle size distribution in the products and high efficiency to obtain nano or microparticles. The optimization of the process on the yield and quality of obtained particles properties depend mainly on the operational conditions such as pressure, temperature, and concentration of the bioactives solution, in terms of extract and polymer. Thus, this chapter provides some insights about the fundamentals and effects of operational conditions of these SC–CO$_2$ antisolvent processes.

1.1 Introduction

A supercritical fluid (SCF) is any fluid that is at a pressure and temperature condition above its critical point. The main interest of this type of fluid is that they have special physical properties, specifically liquid-like density, gas-like diffusivity and viscosity, and zero surface tension (Tabernero et al. 2016).

Conventional techniques for particle formation have been proposed in the literature (spray-drying, jet milling, liquid antisolvent precipitation, solvent evaporation, emulsification, and lyophilization. However, these methods suffer from many drawbacks, mainly lack control over particle morphology, particle size, and particle size distribution (PSD), difficulty in the elimination of the solvents used, and possible degradation due to high temperatures employed (Wang et al. 2013).

Particle engineering for solute delivery needs methodologies that can provide control of particle size and polymorphic purity. Particles should be around 0.1–0.3 μm for intravenous delivery, 1–5 μm for inhalation delivery, and 0.1–100 μm

© The Author(s), under exclusive license to Springer Nature Switzerland AG 2020
D. T. Santos et al., *Supercritical Fluid Biorefining*,
SpringerBriefs in Applied Sciences and Technology,
https://doi.org/10.1007/978-3-030-47055-5_1

for oral delivery (York et al. 2004). Small size also implies a greater percentage of solute absorbed by the human body and a reduction of the doses number. On the other hand, crystallinity affects the physical and chemical stability, whereas organic and inorganic impurities indicate toxicity (Shekunov and York 2000). Supercritical fluids (SCFs) technologies can provide a particle size less than 1 μm, offering in addition a clean technology (Tabernero et al. 2012).

In the industry, a range of applications have been reported and some of the challenging areas of particle design which have been addressed by SCF processing are: (i) Particle morphology control; (ii) Stabilization of therapeutic agents derived from biotechnology, (iii) Polymorph screening; (iv) Improved performance of poorly soluble compounds; (v) Inhaled therapy; (vi) Taste masking; (vii) Crystallization seeds; (viii) Low residual solvent in SCF processed particles (Crystec Pharma 2013).

A fluid is defined as supercritical when its temperature and pressure exceed critical values (see Fig. 1.1). Its solvency power is enhanced due to its higher density, which is very similar to those of liquids (0.1–0.9 g.cm^{-3} at 7.5–50 MPa). Furthermore, these fluids have the main characteristics of gases such as low viscosity, large diffusivity, and small surface tension, which are favorable characteristics for a variety of processes (Skala and Orlovic 2006).

For the industry, carbon dioxide is particularly attractive. The main advantages of CO_2 are its non-toxicity, non-flammability, its relatively low critical temperature, and pressure ($T_c = 304.2$ K, $P_c = 7.38$ MPa), it is inexpensive, recyclable, environmentally acceptable, it has GRAS (Generally Regarded as Safe) status at regulatory agencies, it does not cause oxidation and its solvation properties can be customized by the addition of cosolvents and/or changing the operating conditions, it can be completely separated from the final product by expansion, and then liquefied for recycling purposes.

Fig. 1.1 Pressure–temperature phase diagram of a single substance

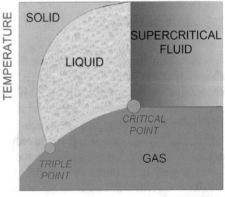

1.2 Supercritical Antisolvent (SAS) Precipitation Fundamentals

Supercritical antisolvent (SAS) precipitation process is based on two prerequisites: the solvent and the antisolvent (CO_2) must be completely miscible at process conditions; the solute must be insoluble in the mixture solvent/antisolvent (Prosapio et al. 2018).

Supercritical antisolvent precipitation is also known as GAS (gas antisolvent), PCA (precipitation by compressed antisolvent), ASES (aerosol solvent extraction system), SEDS (solution-enhanced dispersion by supercritical fluids), and SAS (Supercritical Antisolvent) (Reverchon 1999). These processes are essentially the same, with differences in the feed mode of the organic solvent and antisolvent, which can be co-current or counter-current, depending on the type of injector used, and can use batch (GAS) or semi-continuous (SAS) modes (Adami et al. 2008).

Modifications of the SAS process have been proposed by different authors, such as the use of a coaxial nozzle (solution-enhanced dispersion of supercritical fluids, SEDS); for avoiding solubility drawbacks to micronize proteins or sugars or the use of a tube as a nozzle (atomized rapid injection solvent extraction, ARISE), to avoid problems concerning the blockage of capillary nozzles due to the freezing process in the depressurization. With the aim of increasing the mass transfer in the SAS technique, it is possible to use ultrasound in the process called SAS precipitation with enhanced mass transfer (Tabernero et al. 2016).

In Gas Antisolvent (GAS) and Supercritical Antisolvent (SAS), SC–CO_2 acts as an antisolvent (Bertucco et al. 1998), The GAS method consists of the addition of SC–CO_2 into a bioactive solution (BS) formed by organic solvent and the solute of interest and dissolves the solvent from the solution. The SAS technique is based on the same principles as the GAS but in a continuous or semi-continuous operation. Supercritical Assisted Atomization (SAA) in which SC–CO_2 acts as a propeller (Reverchon et al. 2003), Rapid Expansion of Supercritical Solutions (RESS) in which SC–CO_2 acts as a solvent (Debenedetti et al. 1993), and Particles from Gas Saturated Solutions (PGSS) in which SC–CO_2 acts as a solute (Weidner et al. 1995).

At the same time within each of these techniques, it is possible to identify subvariants with their own characteristics. In Fig. 1.2, we presented a schematic diagram of these processes for a better understanding of their particularities and differences. The SAS technique was highlighted in order to introduce the subvariant SAF technique.

1.3 SAS/SAF Operational Processing

In the SAS process, the solute of interest is dissolved or suspended in a suitable organic solvent. This organic solvent is then contacted with an antisolvent (supercritical (SC) CO_2) with a low affinity for solutes and appreciable mutual solubility with the organic phase. Nucleation and growth of crystals from this solute/organic

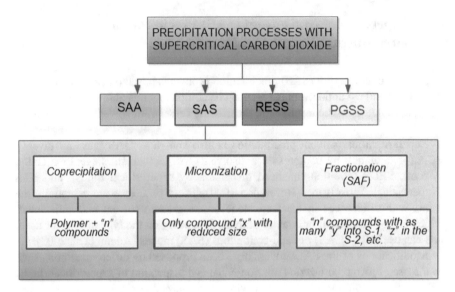

Fig. 1.2 Schematic diagram of the main supercritical fluid-based precipitation processes

solvent/SC–CO_2 system is governed by two mechanisms: diffusion of the antisolvent inside the organic phase and the evaporation of the organic solvent into the antisolvent phase. Diffusion phenomenon increases the volume of the solvent, reduces the density of the solvent, thus decreasing the solvating power of the solvent, and precipitates the solute (Kalani et al. 2011). Therefore, high supersaturation can be achieved (Subra and Jestin 1999). This process typically operates at moderate pressures (9–15 MPa). When the aim is to encapsulate, the solution is prepared using the solute, normally a biodegradable polymer and an organic solvent capable of dissolving both compounds. A schematic representation of SAS precipitation process is presented in Fig. 1.3. This process consists of a precipitation vessel in which CO_2 (supercritical for SAS, or subcritical for PCA) is first pumped inside the high-pressure vessel until the system reaches the fixed pressure and temperature, then, the organic solution is sprayed through a nozzle into the SCF bulk, determining the formation of the particles that are collected on a filter at the bottom of the vessel (Pasquali and Bettini 2008). A wide range of precipitation vessel volumes has already been used for solute micronization. The precipitation vessels used in pilot scale have volumes of 2 L (Montes et al. 2012) and 5 L (Reverchon et al. 2003).

The main advantages of this process are (Kalani et al. 2011; Majerik et al. 2007, Yeo and Kiran 2005; RSC 2005): (i) it is possible to produce narrower particles size distribution by controlling the operating condition; (ii) the supercritical fluid is easily removed from the system by reducing the pressure; (iii) the process can take place at near ambient temperatures, thus avoiding thermal degradation of the particles; (iv) the process can prepare particles with high polymorphic purity, enhanced dissolution rate, and acceptable residual solvent; (v) this process is adaptable for continuous operations, and this property is very important for the large-scale mass production of

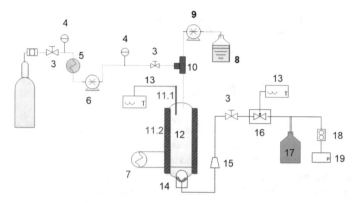

Fig. 1.3 Schematic diagram of the SAS/SAF apparatus. 1—CO$_2$ cylinder; 2—CO$_2$ filter; 3—blocking valves; 4—manometers; 5—cooling bath; 6—CO$_2$ pump; 7—heating bath; 8—solution (solute/solvent) preservoir; 9—HPLC pump; 10—nozzle; 11.1—thermocouple; 11.2—heating jacket; 12—precipitation vessel; 13—temperature controllers; 14—filter of particles collecting; 15—line filter; 16—micrometering valve with a heating system; 17—flask collector; 18—glass float rotameter; 19—flow totalizer

fine particles; (vi) the process is an enclosed system, which vastly reduces the risk of contamination; (vii) being a single-step, self-contained process, it tends to use less solvent than conventional approaches; additionally, it allows solvents recycling.

Different supercritical carbon dioxide (SC–CO$_2$) based techniques have been proposed to micronize several kinds of compounds of interest to the food and non-food industries, like plant extracts, superconductors, explosives, polymers, and biopolymers (Prosapio et al. 2015). Besides micronization, SC–CO$_2$ techniques have been used for the fractionation (supercritical antisolvent fractionation—SAF) of complex matrix, like plant extracts (Guamán-Balcázar et al. 2017) and for the improvement of the component of interest with the use of polymers as wall material (Adami et al. 2017). The use of supercritical CO$_2$ (SC–CO$_2$) antisolvent for micronization, coprecipitation, and fractionation of high-value products for biorefining of plant matrices into marketable products has been a promising and increasing research topic.

1.4 Effect of Operational Conditions in SAS/SAF Process

Since these three past decades, many research (Reverchon et al. 2003; Thiering et al. 2001) and review (York 1999; Knez and Weidner 2003; Temelli 2018; Prosapio et al. 2018) papers have been published in order to solve shortcomings of conventional techniques of particle design and purification of bioactive constituents by investigating the potential of SC–CO$_2$-antisolvent-based techniques

It is more likely to accurately control product characteristics using SAS/SAF because of the large number of operating parameters and conditions (pressure, temperature, nature of the phases in presence, compositions, flow rates, introduction modes, mixing conditions, etc.), which can be varied in order to obtain an end-product with specific properties (Badens et al. 2018).

1.4.1 Temperature and Pressure

The density of SC–CO$_2$, which is influenced by pressure and temperature, affects mass transfer between organic solvent and SC–CO$_2$ during precipitation. Increase in density enhances the molecular interaction, and, thus, the solubility of processed compounds (Kalani et al. 2011). As the pressure increases particle size decreases. Fast mixing and mass transfer at higher CO$_2$ densities lead to smaller particles. However, as the density increases diffusivity of the organic solvent decreases slowing down the mass transfer and mixing which results in larger particles (Uzun et al. 2011).

Expansion of solutions to pressure conditions above ambient, and thereby at lower levels of supersaturation, can result in agglomeration of particles, whereas increased supersaturation during expansion leads to extremely rapid nucleation rates and micron- and submicron-sized particle (York 1999).

Increasing the temperature reduces the solubility and thus enhances the maximum supersaturation, so that smaller particles are obtained. Also, higher temperature reduces the drying time and thus there is rapid removal of the residual solvent; therefore, more spherical particles are formed (Kalani et al. 2011).

In SAF reported to eucalyptus extract at a constant temperature with increasing pressure lead to a slight decrease in both particle size and particle size distribution. At higher pressures, the solubility of the compounds in CO$_2$ is higher, but the solubility in the organic solvent is lower so supersaturation is favored leading to smaller particle size (Chinnarasu et al. 2015).

The phase behavior of the mixtures in the precipitator, during the precipitation stage, is one of the crucial factors to govern the product's morphology and particle size distribution (PSD). Small fluctuation of the pressure and temperature during the precipitation may change the phase states of the mixtures in the precipitator. Chang et al. (2008) show that various types of morphologies were obtained as the temperature was increased which included micro-scale porous clusters, irregular submicron, and micron-metric aggregated particles. When the mixtures in the precipitator are in the superheated vapor region, expanded microparticles (ranging from 5 to 20 μm) were formed. When the mixtures in the precipitator were in vapor–liquid coexistence region, sub-micrometric particles (about 200 nm) seriously aggregated to form a dense cake on the surface of the filter were formed.

The operating temperature has a significant effect on polymorph. The variation of temperature results in the formation of different polymorphs. Park et al. (2007) obtained fluconazole anhydrate form I at low temperature (313 K) and anhydrate form II at high temperature (353 K), the increased extraction rate between SC–CO$_2$ and organic solvent, resulting in supersaturation at high temperature, leading to a different polymorph. The variation of pressure during the SAS process may influence the preferred orientation.

The increase in temperature and decrease in pressure resulted in fractionated rosemary extracts with improved polyphenolic content (Visentin et al. 2012). At 313 K, increase in pressure resulted in increase in global yield of solids (GY) in SAF applied to grape extracts (Natolino et al. 2016). The pressure effects on the purification of polyphenols from Schiandrachinensis extracts, resulted in maximum GY of 70.9% at the highest pressure (17 MPa), while the maximum concentration of polyphenols was obtained at the lowest pressure (130 MPa). Authors suggested that both SC–CO_2 and ethanol have high densities at high pressure and thus resulted in higher solubilities to the impurity in the effluent. Such impurity was mixed into the target compounds, which decreased the purities of each fraction but increased the recoveries in solids (Huang et al. 2013).

1.4.2   Solute Concentration

The initial solute solution concentration is related to product yield and process efficacy in the SAS precipitation process (Kim et al. 2012). When the other operational conditions are fixed, the increase of solute concentration causes the increase of the mean particle size and the enlargement of particle size distribution. As the solution concentration is increased the fast reduction of solvent power by the antisolvent CO_2 allows homogeneous particles to nucleate. Beyond a certain point, as the solution concentration is increased, particle growth dominates the nucleation process, which results in larger particles This effect is evident in some papers published on procyanidins (Yang et al. 2011), indomethacin (Varughese et al. 2010), and atorvastatin calcium (Kim et al. 2012). Nevertheless, the contrary effect was reported in SAF increasing of solute concentration resulted in decreased particle size in fractionated mango leaves extracts. A priori the initial concentration of the solution in the SAS process could have two effects on particle size: (1) at higher concentration it is possible to achieve higher supersaturations, thereby reducing particle size and (2) condensation is directly proportional to the concentration of the solution, and the increase in the condensation rate at higher concentrations tends to increase the particle size (Guamán-Balcázar et al. 2017; Martín and Cocero 2004).

The saturation solubility of the solute in mixtures of solvents and antisolvents could be identified as the indirect classification criterion to distinguish between amorphous precipitating and crystallizing SAS systems. Furthermore, it is shown that crystallizing SAS system crystals may be generated from either a supercritical or a liquid phase which is in contrast to the widespread SAS opinion that crystals only can be formed in the liquid phase (Rossmann et al. 2012).

In SAF applied to carotenoids extracts at 20 MPa and 313 K, the use low solute concentrations resulted in an enhancement in the fractionation of zeaxanthin from 40 to 67.4% when compared to of high feed concentration (Liau et al. 2010), on the contrary to those reported with zeaxanthin content on fractionated Nannochloropsisoculata extracts (Cho et al. 2011) and with polar lipids content fractionated egg yolk extracts (Aro et al. 2009).

1.4.3   Nozzle Geometry

Shear stress can be achieved with nozzles, static mixers, or capillaries, and causes a pressure drop that can be in the range of some Pa up to 10 MPa or more, depending on the viscosity of the substances. During the suction phase of the piston pump, the ratio of the gas phase in the nozzle is increased, leading to better disintegration of the liquid and therefore most likely to small droplets. During the pressurization phase of the pump, the ratio of carbon dioxide to liquid is shifted further towards the liquid phase, leading to a worse spray formation and therefore bigger particles (Petermann 2018).

Chang et al. (2008) observed that the smaller diameter of the nozzle produces a higher spray velocity and reduces the droplet size. The authors obtained larger particles with aggregation by using 508 μm capillary tube and by using 254 μm capillary tube, smaller spherical primary nanoparticles with less agglomeration were produced. It was suggested that the smaller inside diameters of capillary tube could effectively atomize the solution to form the finer droplets. But another research has shown that the effect of the nozzle diameter is not very significant (Chen et al. 2010). Thus, more research is needed to fully understand the influence of nozzle geometry on particle formation by SAS process.

Effects of nozzle diameter in the fractionation of rosemary extracts resulted in an increase in global yield when using the 0.130 mm diameter nozzle at 40 °C and 10 MPa, and no noteworthy change at 50 °C and 10 MPa, probably to mechanical limitations in the recovery of the particles (Visentin et al. 2012).

Supercritical Antisolvent Fractionation for obtaining nano-sized curcumin particles was done by inserting its bioactives solution using a capillary nozzle with internal diameter of 100 μm, employing the conditions of 14 MPa, 313 K, CO_2 flow rate at 40 g/min, solution flow rate at 0.5 mL/min, and solution concentration at 0.5% (Anwar et al. 2015).

1.4.4 CO_2 and Solution Flow Rates

Increasing the ratio of CO_2 flow rate to the organic solution flow rate reduces the particle size (Kalani et al. 2011). At lower solution flow rate, the supersaturation of the solutes occurs rather slowly. Therefore, the precipitation delays and nucleation dominate growth. Besides, the increase in flow rate can enhance the viscosity and surface tension of the solution. Thus, microparticles have the tendency to aggregate at higher flow rate (Wang et al. 2013).

In SAF, increasing in CO_2 flow rate resulted in decreasing in the global yield of solids and polyphenols from *Schisandra chinensis* extracts (Huang et al. 2013). In addition, increasing in total lignans from fractionated flaxseed extracts were attributed to increases in CO_2 flowrate (Perretti et al. 2013).

1.4.5 Type of Organic Solvent

The choice of solvent influences crystallization of the polymorphic forms. It was apparent that methanol as a hydrogen bond donor possessed a much greater ability to stabilize different forms of sulfathiazole than acetone (Shariati and Peters 2003).

Among the various choices of solvents, dichloromethane is a common choice for SAS experiments because of its dissolution of a wide number of compounds or materials and good miscibility with SC–CO_2 at low temperatures and pressures (Varughese et al. 2010).

The different morphologies observed using different solvents are attributed to the differences in chemical properties between the organic solvents and carbon dioxide, and the properties of the solvents. Essentially, CO_2 behaves as a nonpolar solvent and is most soluble in the least polar solvent. Thus, at a given pressure, the level of volumetric expansion is greater for a less polar solvent than for a more polar solvent, and small spherical nanoparticles were produced from solvents that have large volume expansions.

Particles with various morphologies, from spheroidal to agglomerated and particle size ranging between 0.04 and 0.12 μm were produced using N,N-dimethylformamide DMF as a solvent for the coprecipitation of lycopene (Nerome et al. 2013). Reverchon (1999) reported that very small (100–200 nm) nanoparticles were produced from solvents that have large volume expansions.

The interaction of the liquid solvent with SC–CO_2 can induce an increase in the solubility of the compounds of interest in SC–CO_2. In this case, the liquid can act as a cosolvent from the point of view of the solid solubilization. When this phenomenon occurs, the part of the solute retained in the fluid phase does not precipitate during SAF and is recovered in the precipitation or separation vessel (Martín et al. 2011).

Morphology and particle size are dominated by the mechanisms of organic solvent evaporation into the antisolvent phase and antisolvent diffusion into the organic solvent. Therefore, the properties of organic solvents such as polarity, boiling point, and solubility parameter are important variables in controlling particle formation using the SAS process (Kim et al. 2007). The choice of organic solvent is also important for the control of crystallization of solute polymorphs because the type of organic solvent may affect supersaturation during the SAS process (Park et al. 2007).

In SAF ethanol/water extracts a larger amount of polar, high molecular mass compounds from the solubilized extracts as well as glycosylated flavonoids, which are not expected to be soluble in CO_2 + ethanol (Catchpole et al. 2004). Several analytical-scale studies have shown that polar lipids can be dissolved if small amounts of cosolvents are added to SC–CO_2. When only GRAS-grade solvents are included, ethanol is the most used for the isolation of polar lipids from egg yolk (Aro et al. 2009). The appropriate solvent may improve fractionation of compounds of interest.

1.5   Conclusions

From a scientific point of view, particle design using the SAS precipitation and SAF process are attractive options due to the possibility of obtaining particles with controlled particle size and morphology, narrow size distribution, and acceptable residual organic solvent content.

The main advantages of these processes are (i) the processes can take place at near ambient temperatures, thus avoiding thermal degradation of the processed solutes; (ii) they are adaptable for continuous operations being possible large-scale mass production of fine particles; (iii) they allow solvent (CO_2 and organic solvent) recycling. Few reports have compared the particles obtained by both processes. On the other hand, several studies have demonstrated the same trend: solute processing by SAS or SAF improves its dissolution rate an concentration, which is one key point for the commercial development of supercritical fluid biorefineries using supercritical CO_2 as an antisolvent for micronization, coprecipitation, and fractionation of high-value products for biorefining of plant matrices into marketable products.

Acknowledgements   Diego T. Santos thanks CNPq (processes 401109/2017-8; 150745/2017-6) for the post-doctoral fellowship. Ricardo A. C. Torres thanks Capes for their doctorate assistantship. Ádina L. Santana thanks CAPES (1764130) for the post-doctoral fellowship. M. Angela A. Meireles thanks CNPq for the productivity grant (302423/2015-0). The authors acknowledge the financial support from CNPq (process 486780/2012-0) and FAPESP (processes 2012/10685-8; 2015/13299-0).

References

Adami R, Reverchon E, Järvenpää E, Huopalahti R (2008) Supercritical AntiSolvent micronization of nalmefene HCl on laboratory and pilot scale. Powder Technol 182:105–112

Adami R, Di Capua A, Reverchon E (2017) Supercritical assisted atomization for the production of curcumin-biopolymer microspheres. Powder Technol 305:455–461

Anwar M, Ahmad I, Warsi MH, Mohapatra S, Ahmad N, Akhter S, Ali A, Ahmad FJ (2015) Experimental investigation and oral bioavailability enhancement of nano-sized curcumin by using supercritical anti-solvent process. Eur J Pharm Biopharm 96:162–172

Aro H, Järvenpää EP, Könkö K, Sihvonen M, Hietaniemi V, Huopalahti R (2009) Isolation and purification of egg yolk phospholipids using liquid extraction and pilot-scale supercritical fluid techniques. Eur Food Res Technol 228:857–863

Badens E, Masmoudi Y, Mouahid A, Crampon C (2018) Current situation and perspectives in drug formulation by using supercritical fluid technology. J Supercrit Fluids 134:274–283

Bertucco A, Lora M, Kikic I (1998) Fractional crystallization by gas antisolvent technique: theory and experiments. AIChE J 44:2149–2158

Catchpole OJ, Grey J, Mitchell K, Lan J (2004) Supercritical antisolvent fractionation of propolis tincture. J Supercrit Fluids 29:97–106

Chang SC, Lee MJ, Lin HM (2008) Role of phase behavior in micronization of lysozyme via a supercritical anti-solvent process. Chem Eng J 139:416–425

Chen YM, Tang M, Chen YP (2010) Recrystallization and micronization of sulfathiazole by applying the supercritical antisolvent technology. Chem Eng J 165:358–364

Chinnarasu C, Montes A, Fernandez-Ponce MT, Casas L, Mantell C, Pereyra C, de la Ossa EJM, Pattabhi S (2015) Natural antioxidant fine particles recovery from *Eucalyptus globulus* leaves using supercritical carbon dioxide assisted processes. J Supercrit Fluids 101:161–169

Cho Y-C, Cheng J-H, Hsu S-L, Hong S-E, Lee T-M, Chang C-MJ (2011) Supercritical carbon dioxide anti-solvent precipitation of anti-oxidative zeaxanthin highly recovered by elution chromatography from *Nannochloropsis oculata*. Sep Purif Technol 78:274–280

Crystec Pharma. Supercritical fluid processing for controlled particle formation. Available at: http://www.crystecpharma.com/index.php?id=23. Accessed: Apr 2013

Debenedetti PG, Tom JW, Yeo SD (1993) Rapid expansion of supercritical solutions (RESS): fundamentals and applications. Fluid Phase Equilib 82:311–318

Guamán-Balcázar MC, Montes A, Pereyra C, de la Ossa EM (2017) Precipitation of mango leaves antioxidants by supercritical antisolvent process. J Supercrit Fluids 128:218–226

Huang T-L, Lin JC-T, Chyau C-C, Lin K-L, Chang C-MJ (2013) Purification of lignans from *Schisandra chinensis* fruit by using column fractionation and supercritical antisolvent precipitation. J Chromatogr A 1282:27–37

Kalani M, Yunus R, Abdullah N (2011) Optimizing supercritical antisolvent process parameters to minimize the particle size of paracetamol nanoencapsulated in L- polylactide. Int J Nanomed 6:1101–1105

Kim MS, Lee S, Park JS, Woo JS, Hwang SJ (2007) Micronization of cilostazol using supercritical antisolvent (SAS) process: effect of process parameters. PowderTechnology 177:64–70

Kim MS, Song HS, Park HJ, Hwang SJ (2012) Effect of solvent type on the nanoparticle formation of atorvastatin calcium by the supercritical antisolvent process. Chem Pharm Bull 60:543–547

Knez Z, Weidner E (2003) Particles formation and particle design using supercritical fluids. Curr Opin Solid State Mater Sci 7:353–361

Liau BC, Shen CT, Liang FP, Hong SE, Hsu SL, Jong TT, Chang CMJ (2010) Supercritical fluids extraction and anti-solvent purification of carotenoids from microalgae and associated bioactivity. J Supercrit Fluids 55:169–175

Majerik V, Badens GCE, Horváth G, Szokonya L, Bosc N, Teillaud E (2007) Bioavailability enhancement of an active substance by supercritical antisolvent precipitation. J Supercrit Fluids 40:101–110

Martín A, Cocero MJ (2004) Numerical modeling of jet hydrodynamics, mass transfer, and crystallization kinetics in the supercritical antisolvent (SAS) process. J Supercrit Fluids 32:203–219

Montes A, Gordillo MD, Pereyra C, Ossa EJM (2012) Polymer and ampicillin co- precipitation by supercritical antisolvent process. J Supercrit Fluids 63:92–98

Natolino A, Da Porto C, Rodríguez-Rojo S, Moreno T, Cocero MJ (2016) Supercritical antisolvent precipitation of polyphenols from grape marc extract. J Supercrit Fluids 118:54–63

Nerome H, Machmudah S, Wahyudiono Fukuzato R, Higashiura T, Youn Y-S, Lee Y-W, Goto M (2013) Nanoparticle formation of lycopene/β-cyclodextrin inclusion complex using supercritical antisolvent precipitation. J Supercrit Fluids 83:97–103

Park HJ, Kim MS, Lee S, Kim JS, Woo JS, Park JS, Hwang SJ (2007) Recrystallization of fluconazole using the supercritical antisolvent (SAS) process. Int J Pharm 328:152–160

Pasquali I, Bettini (2008) Are pharmaceutics really going supercritical? Int J Pharm 364:176–187

Perretti G, Virgili C, Troilo A, Marconi O, Regnicoli GF, Fantozzi P (2013) Supercritical antisolvent fractionation of lignans from the ethanol extract of flaxseed. J Supercrit Fluids 75:94–100

Petermann M (2018) Supercritical fluid-assisted sprays for particle generation. J Supercrit Fluids 134:234–243

Prosapio V, De Marco I, Reverchon E (2018) Supercritical antisolvent coprecipitation mechanisms. J Supercrit Fluids 138:247–258

Prosapio V, Reverchon E, De Marco I (2015) Control of powders morphology in the supercritical antisolvent technique using solvent mixtures. Chem Eng Trans 43:763–768

Reverchon E, Caputo G, De Marco I (2003) Role of phase behavior and atomization in the supercritical antisolvent precipitation. Ind Eng Chem Res 42:6406–6414

Reverchon E (1999) Supercritical antisolvent precipitation of micro- and nano-particles. J Supercrit Fluids 15:1–21

Rossmann M, Braeuer A, Leipertz A, Schluecker E (2012) Supersaturation as criterion for the appearance of amorphous particle precipitation or crystallization in the supercritical antisolvent (SAS) process. In: 10th international symposium on supercritical fluids (ISSF). ISSF 2012 proceedings. RSC—advancing the chemical sciences. Supercritical fluids: realising potential. Available at: http://www.rsc.org/chemistryworld/Issues/2005/February/supercriticalfluids.asp Accessed Apr 2013

Shariati A, Peters CJ (2003) Recent developments in particle design using supercritical fluids. Curr Opin Solid State Mater Sci 7:371–383

Shekunov BY, York P (2000) Crystallization processes in pharmaceutical technology and drug delivery design. J Cryst Growth 211:122–136

Skala D, Orlovic A (2006) Particle production using supercritical fluids. In: Hsu J-P, Spasic AM (eds) Micro-, nano-, and Atto-Engineering. Taylor & Francis Group

Subra P, Jestin P (1999) Powders elaboration in supercritical media: comparison with conventional routes. Powder Technol 103:2–9

Tabernero A, González-Garcinuño Á, Galán Miguel A, Martín del Valle Eva M (2016) Survey of supercritical fluid techniques for producing drug delivery systems for a potential use in cancer therapy. Rev Chem Eng 32

Tabernero A, Valle EMM, Galán MA (2012) Supercritical fluids for pharmaceutical particle engineering: methods, basic fundamentals and modeling. Chem Eng Process 60:9–25

Temelli F (2018) Perspectives on the use of supercritical particle formation technologies for food ingredients. J Supercrit Fluids 134:244–251

Thiering R, Dehghani F, Foster NR (2001) Current issues relating to anti-solvent micronisation techniques and their extension to industrial scales. J Supercrit Fluids 21:159–177

Uzun IN, Sipahigil O, Dinçer S (2011) Coprecipitation of Cefuroxime Axetil–PVP composite microparticles by batch supercritical antisolvent process. J Supercrit Fluids 55:1059–1106

Varughese P, Li J, Wang W, Winstead D (2010) Supercritical antisolvent processing of γ-Indomethacin: effects of solvent, concentration, pressure and temperature on SAS processed Indomethacin. Powder Technol 201:64–69

Visentin A, Rodríguez-Rojo S, Navarrete A, Maestri D, Cocero MJ (2012) Precipitation and encapsulation of rosemary antioxidants by supercritical antisolvent process. J Food Eng 109:9–15

Wang W, Liu G, Wu J, Jiang Y (2013) Co-precipitation of 10-hydroxycamptothecin and poly(l-lactic acid) by supercritical CO_2 anti-solvent process using dichloromethane/ethanol co-solvent. J Supercrit Fluids 74:137–144

Weidner E, Knez Z, Novak Z (1995) Int Pat Publ WO 95/21688

Yang, L, Huang JM, Zu YG, Ma CH, Wang H, Sun XW, Sun Z (2011) Preparation and radical scavenging activities of polymeric procyanidins nanoparticles by a supercritical antisolvent (SAS) process. Food Chem 128:1152–1159

Yeo SD, Kiran E (2005) Formation of polymer particles with supercritical fluids: a review. J Supercrit Fluids 34:287–308

York P (1999) Strategies for particle design using supercritical fluid technologies. Pharm Sci Technol Today 2:430–440

York P, Kompella UB, Shekunov BY (2004) Supercritical fluid technology for drug product development. Marcel-Dekker, New York

Chapter 2
Supercritical Fluid Biorefining Using Supercritical CO$_2$ as an Antisolvent for Micronization, Coprecipitation, and Fractionation: Recent Applications

Abstract This chapter discusses the recent (2013–2018) applications on the use of supercritical CO$_2$ (SC–CO$_2$) antisolvent for micronization, coprecipitation, and fractionation of high-value products for the food, cosmetic, and pharmaceutical industries, most focusing on the applications of integrated techniques on the biorefining of plant matrices into marketable products using supercritical carbon dioxide as an antisolvent. The concept of the biorefinery is defined as sustainable processing of feedstocks for bioenergy and biochemical purposes resulting in various marketable products, such as natural dyes, antioxidants, proteins for food and feed, lipids for biodiesel, and carbohydrates as feedstock for bioethanol production

2.1 Introduction

Recently, supercritical carbon dioxide (SC–CO$_2$) has been indicated to be utilized as a key green solvent in biorefining, for the extraction and recovery of products with active principles indigenous to various plant matrices, such as lipids from algae, waxes from date palm leaves, polyphenols from buckwheat flowers and hulls, volatile oils, pigments, starch, and for particle devices preparation from turmeric, among others (Santana et al. 2019).

Rather than a stand-alone technology, the utilization of supercritical technology, i.e., processes which uses supercritical fluids and pressurized liquids, as part of an integrated biorefinery is a recent trend for obtaining of nutraceuticals and has been shown to have a positive effect on the efficiency processing of different solid matrix (Albarelli et al. 2016).

This review examines the current research scenario that justifies the biorefining of crude and waste plant matrices with the use of supercritical CO$_2$ as an antisolvent for micronization, coprecipitation, and fractionation. The period reviewed ranges from 2013 to 2018.

2.2 Applications for SAS

Supercritical Antisolvent (SAS) allows for the precipitation of bioactive constituents and biopolymers in micro- and nanometer size in a wide range of industrial applications while guaranteeing the physical and chemical integrity of such materials (Almeida et al. 2016). Recently, this technique has been used for the processing of plant extracts for micronization (Kurniawansyah et al. 2017; Montes et al. 2016c), coprecipitation (Machado et al. 2018) and fractionation, i.e., SAF process (Chinnarasu et al 2015). Table 2.1 shows the recent (2013–2018) reports on the use of supercritical CO_2 (SC–CO_2) antisolvent for micronization, coprecipitation, and fractionation.

2.3 Applications for SAF

The process parameters that influence the performance of a SAF process are temperature, pressure, flow of antisolvent and BS, type of organic solvent, characteristics of the nozzle, the concentration of the bioactive compound in the BS, time of processing and characteristic of the vessel (Torres et al. 2016). The purification progress of several products using SAF is represented by the yield of bioactives, which is the ratio between the bioactive quantified in the fractionated extracts (particles) divided by the same bioactive presented in the crude extract (Table 2.2).

2.4 Conclusions

The total utilization of plant matrices using multistep procedures for biorefining purposes is a feasible alternative for the obtaining of various applications in food, pharmacy, biofuels, and energy production.

Recent trends on supercritical technology focused on the development of biorefinery concepts applied to waste-biomass constituents shows that this technology is efficient when compared to conventional ones, in terms of extraction yield and recovery of active constituents. New perspectives on how biomasses can be better valorized with the aid of using CO_2 as an antisolvent was shown in this chapter, indicating that SAS/SAF can be effectively used in biorefinery concepts aiming full use of biomass prior to its chemical conversion.

Acknowledgements Diego T. Santos thanks CNPq (processes 401109/2017-8; 150745/2017-6) for the post-doctoral fellowship. Ricardo A. C. Torres thanks Capes for their doctorate assistantship. Ádina L. Santana thanks CAPES (1764130) for the post-doctoral fellowship. M. Angela A. Meireles thanks CNPq for the productivity grant (302423/2015-0). The authors acknowledge the financial support from CNPq (process 486780/2012-0) and FAPESP (processes 2012/10685-8; 2015/13299-0).

Table 2.1 Reported applications of SAS/SAF precipitation

Substance	Process	Solvent/antisolvent	Operational parameters	Results	References
Anthocyanins	Coprecipitation	Ethanol/SC CO_2	Pressure: 12.5 MPa Temperature: 302–318 K Wall material polyvinylpyrrolidone PVP Polymer concentration: 0.5% Actives concentration: – Solution flow rate: 0.060–0.6 $L \cdot h^{-1}$ CO_2 flow rate: 2.0 $kg \cdot h^{-1}$	Particle size: 9.7–206.32 μm Morphology: spherulite Structure: – Polymorphic nature: – Precipitation yield: 38.39–76.05% Coprecipitation efficiency: –	Machado et al. (2018)
Terpenes from eucalyptus leaves extract	SAF	Ethanol/SC–CO_2	Pressure: 10–20 MPa Temperature: 308–323 K Actives concentration: – Solution flow rate: 0.12–0.3 $L \cdot h^{-1}$ CO_2 flow rate: 0.6–1.2 $kg \cdot h^{-1}$	Particle size: 0.27–0.92 μm Morphology: spherical Structure: – Polymorphic nature: – Precipitation yield: – Precipitation efficiency: –	Chinnarasu et al. (2015)
Catechins and caffeine from tea leaves	SAF	Ethyl lactate/SC CO_2 Ethanol/SC CO_2	Pressure: 10–30 MPa Temperature: 323–353 K The CO_2/ethyl lactate extract flow ratio: 40 L/L CO_2/ethanol extract flow ratio: 20 L/L	Particle size: ≈100 μm Morphology: irregular Structure: quasi-amorphous Polymorphic nature: – Precipitation yield: 32.8–64% Precipitation efficiency: 1.32–68%	Villanueva-Bermejo et al. (2015)

(continued)

Table 2.1 (continued)

Substance	Process	Solvent/antisolvent	Operational parameters	Results	References
Curcumin	Micronization	Acetone/Compressed-SCCO$_2$ Ethanol/Compressed-SC CO$_2$ Acetone-Ethanol/Compressed-SCCO$_2$	Pressure: 9.5–10 MPa Temperature: 298–313 K Actives concentration: – Solution flow rate: 0.06–0.6 mL·h^{-1} CO$_2$ flow rate: 1.2 L·h^{-1}	Particle size: 5–7 μm Morphology: crystal form I, II and II Structure: – Polymorphic nature: – Precipitation yield: % Precipitation efficiency: 11.5–60.2%	Kurniawansyah et al. (2017)
Curcumin	Micronization	Acetone/SC–CO$_2$ Methanol/SC–CO$_2$	Pressure: 8–16 MPa Temperature: 308–323 K Actives concentration: – Solution flow rate: 0.03–0.12 L·h^{-1} CO$_2$ flow rate: 1.8–2.7 kg·h^{-1}	Particle size: 8–980 μm Morphology: irregular plates Structure: – Polymorphic nature: – Precipitation yield: 67.2–79.7% Precipitation efficiency: – Size reduction: 90.83%	Anwar et al. (2015)
Curcumin	Coprecipitation	Ethanol/SC CO$_2$	Pressure: 9.3–9.9 MPa Temperature: 353 K Wall material: Polyvinylpyrrolidone (PVP) Actives/polymer concentration: 11.25–15 mg·mL^{-1} Solution flow rate: – CO$_2$ flow rate: –	Particle size: 0.94–2.3 μm Morphology: spherical Structure: amorphous Polymorphic nature: – Precipitation yield: – Coprecipitation efficiency: 97–100%	Adami et al. (2017)

(continued)

Table 2.1 (continued)

Substance	Process	Solvent/antisolvent	Operational parameters	Results	References
Curcumin	Coprecipitation	Ethanol/SC CO_2	Pressure: 10–14 MPa Temperature: 308–313 K Wall material: Eudragit® L100, Pluronic® F127, Polyvinylpyrrolidone (PVP) Polymer concentration: – Actives concentration:– Solution flow rate: 0.06–0.12 $L \cdot h^{-1}$ CO_2 flow rate: 2 $kg \cdot h^{-1}$	Particle size: 5.67–8.71 μm Morphology: spherical Structure: – Polymorphic nature: – Precipitation yield: – Coprecipitation efficiency: –	Arango-Ruiz et al. (2018)
Nano-curcumin	Coprecipitation	Acetone-Ethanol/SC–CO_2 Ethyl acetate-Ethanol/SC–CO_2	Pressure: 8–10 MPa Temperature: 296–311 K Wall material: poly(lactic-co–glycolic acid) (PLGA) Actives concentration: – Solution flow rate: 0.03–0.15 $L \cdot h^{-1}$ CO_2 flow rate: 0.6–2.7 $L \cdot h^{-1}$	Particle size: 63–1680 nm Morphology: irregular crystalline Structure: amorphous Polymorphic nature: – Precipitation yield: 12–51% Coprecipitation efficiency: 4–38%	Zabihi et al. (2014)
Curcuminoids from turmeric extracts	SAF	Ethanol/SC–CO_2	Pressure: 10–20 MPa Temperature: 313–333 K Actives concentration: – Solution flow rate: 0.030 $L \cdot h^{-1}$ CO_2 flow rate: 0.5–0.8 $kg \cdot h^{-1}$	Particle size: 111–829 μm Morphology: irregular plates Structure: – Polymorphic nature: – Precipitation yield: 59–97% Precipitation efficiency: 60–97%	Osorio-Tobón et al. (2016)

(continued)

Table 2.1 (continued)

Substance	Process	Solvent/antisolvent	Operational parameters	Results	References
Ellagic acid	Coprecipitation	1-methyl-2-pyrrolidone (NMP)/ SC–CO_2	Pressure: 8–20 MPa Temperature: 313–323 K Wall material: Eudragit Actives concentration: 10–35 mg·mL^{-1} Polymer concentration: 10 mg·mL^{-1} Solution flow rate: 3 mL·min^{-1} CO_2 flow rate: 0.060–0.180 kg·h^{-1}	Particle size: 300–1060 nm Morphology: crystalline, spherical Structure: – Polymorphic nature: – Precipitation yield: – Encapsulation efficiency: 21.44–49.13%	Montes et al. (2016a)
Fucoxanthin	Micronization	Ethyl ether/ SC–CO_2	Pressure: 11 MPa Temperature: 253–313 K Actives concentration: 10 mg·mL^{-1} Solution flow rate: 0.3–0.5 mL·min^{-1} CO_2 flow rate: 15 L·min^{-1}	Particle size: 882.9–1582.3 nm Morphology: hexagonal plate-like Structure: – Polymorphic nature: – Precipitation yield: 64–97.3% Size reduction: 99.11–99.5%	Chen et al. (2017)
Guaçatonga extracts	Coprecipitation	Ethanol/SC CO_2	Pressure: 10–15 MPa Temperature: 308–323 K Wall material: Pluronic F127 Actives concentration: 7.9 mg·mL^{-1} Polymer concentration: 24 mg·mL^{-1} Solution flow rate: 0.06 L·h^{-1} CO_2 flow rate: 505.8 L·h^{-1}	Particle size: 1.61–187 µm Morphology: crystalline irregular Structure: – Polymorphic nature: – Precipitation yield: – Precipitation efficiency: –	Benelli et al. (2014)

(continued)

Table 2.1 (continued)

Substance	Process	Solvent/antisolvent	Operational parameters	Results	References
Lycopene and alpha-Tocopherol	Coprecipitation	N-dimethylformamide and tetrahydrofuran u/SC–CO_2	Pressure: 8–12 MPa Temperature: 308 K Wall material: Lecithin Polymer concentration: 10–20 mg·mL^{-1} Actives concentration: 10–20 mg·mL^{-1} Solution flow rate: 0.2–0.4 ml/min L·h^{-1} CO_2 flow rate: 1.62 kg·h^{-1}	Particle size: 0.87–3.65 µm Morphology: platinum film Structure: – Polymorphic nature: – Precipitation yield: 34.5–45.5% Coprecipitation efficiency: 9.6–29.4%	Cheng et al. (2017)
Mangiferin from mango leaves extracts	Micronization	Acetone: Dimethylsulfoxide (1:3)/SC CO_2	Pressure: 10–15 MPa Temperature: 313–323 K Actives concentration: 7.5–13.5 mg·mL^{-1} Solution flow rate: 0.18–0.24 L·h^{-1} CO_2 flow rate: 0.6–1.8 kg·h^{-1}	Particle size: 0.22–1.44 µm Morphology: spherical Structure: amorphous Polymorphic nature: – Precipitation yield: – Size reduction: 99.59–97.29%	Montes et al. (2016c)
Naringenin	Micronization	Acetone-dichloromethane solution/SC CO_2	Pressure: 8–20 MPa Temperature: 308–323 K Actives concentration: 5–20 mg·mL^{-1} Solution flow rate: 0.06–0.12 kg·h^{-1} CO_2 flow rate: –	Particle size: 0.097–0.585 µm Morphology: crystalline cubic Structure: amorphous Polymorphic nature: – Precipitation yield: – Precipitation efficiency: – Size reduction: –	Miao et al. (2018)

(continued)

Table 2.1 (continued)

Substance	Process	Solvent/antisolvent	Operational parameters	Results	References
Polyphenols from mango by-products	SAF	Ethanol/ SC CO_2	Pressure: 8–15 MPa Temperature: 308–318 K Actives concentration:– Solution flow rate: 60 mL·h^{-1} CO_2 flow rate: –	Particle size: 80.6–227 µm Morphology: spherical Structure: amorphous Polymorphic nature: – Precipitation yield: 63.7–71.6%	Meneses et al. (2015)
Polyphenols from yarrow extracts	SAF	Ethanol/SC CO_2	Pressure: 10–20 MPa Temperature: 308–323 K Actives concentration: 17.9 mg·mL^{-1} Solution flow rate: 0.096 kg·h^{-1} CO_2 flow rate: 3 kg·h^{-1}	Particle size: 250–330 µm Morphology: irregular Structure: quasi-amorphous Polymorphic nature: – Precipitation yield: 6.60–45.5% Precipitation efficiency: 18.9–40%	Villanueva-Bermejo et al. (2017)
Proanthocyanidins from grape marc extract	SAF	Ethanol/SC–CO_2	Pressure: 10–12 MPa Temperature: 313–318 K Actives concentration: – Solution flow rate: 0.12 L·h^{-1} CO_2 flow rate: –	Particle size: 5–10 µm Morphology: irregular Structure: – Polymorphic nature: – Precipitation yield: – Precipitation efficiency: 26–97%	Natolino et al. (2016)
Quercetin from onion peels extract	SAF	Ethanol/SC–CO_2	Pressure: 10–12 MPa Temperature: 313–333 K Actives concentration: – Solution flow rate: 0.024 kg·h^{-1} CO_2 flow rate: 1.020 kg·h^{-1}	Particle size: 119–234 µm Morphology: irregular plates Structure: – Polymorphic nature: – Precipitation yield: 4.1–5.2% Precipitation efficiency: 12.9–26%	Zabot and Meireles (2016)

(continued)

Table 2.1 (continued)

Substance	Process	Solvent/antisolvent	Operational parameters	Results	References
Rutin	SAF	Dimethylsulfoxide/SC CO_2	Pressure: 8–15 MPa Temperature: 313–323 K Actives concentration: 20–36 mg·mL^{-1} Solution flow rate: 3–6 mL·min^{-1} CO_2 flow rate: 0.6–1.8 kg·h^{-1}	Particle size: 0.24–1.9 μm Morphology: spherical Structure: amorphous Polymorphic nature: – Precipitation yield: –	Montes et al. (2016b)
Zeaxanthin palmitates	Coprecipitation	Tetrahydrofuran (THF)/SC–CO_2	Pressure: 15–17.5 MPa Temperature: 328 K Actives concentration: 5 mg·mL^{-1} Solution flow rate: 0.1–0.3 mL·min^{-1} CO_2 flow rate: 900 L·h^{-1}	Particle size: 1525–6839 nm Morphology: irregular Structure: – Polymorphic nature: – Precipitation yield: 48.3–73.7% Precipitation efficiency: 55–74%	Lin et al. (2014)
Polyphenols from mango leaves extract	SAF	Ethanol/ SC CO_2	Pressure: 10–15 MPa Temperature: 308–323 K Actives concentration: 8–40 mg·mL^{-1} Solution flow rate: 0.5–10 mL·min^{-1} CO_2 flow rate: 0.6–2.4 kg·h^{-1}	Particle size: 0.02–30 μm Morphology: spherical Structure: – Polymorphic nature: – Precipitation yield: – Precipitation efficiency: –	Guamán-Balcázar et al. (2017)
Polyphenols from grape waste extract	SAF	Ethanol/ SC CO_2	Pressure: 8–15 MPa Temperature: 308–333 K Actives concentration: 3% Solution flow rate: – CO_2 flow rate: 2.38 kg·h^{-1}	Particle size: 100 nm Morphology: spherical Structure: – Polymorphic nature: – Precipitation yield: 27–62% Precipitation efficiency: 54.58–94.65% Size reduction: 99.97%	Marqués et al. (2013)

(continued)

Table 2.1 (continued)

Substance	Process	Solvent/antisolvent	Operational parameters	Results	References
Passion fruit seed oil	Coprecipitation	Dichloromethane/SC CO_2	Pressure: 9–11 MPa Temperature: 308–318 K Wall material: PLGA (poly(lactic–co–glycolic)acid) Polymer concentration:– Actives concentration:– Solution flow rate:1 mL·min⁻¹ CO_2 flow rate: 1.1–1.6 kg·h⁻¹	Particle size: 721–1498 nm Morphology: spherical Structure:– Polymorphic nature: – Precipitation yield: – Coprecipitation efficiency: 67.8–91%	Dal Magro et al. (2017)
Trans-resveratrol	Coprecipitation	Acetone–Dichloromethane/SC–CO_2	Pressure: 8–15 MPa Temperature: 308 K Wall material: poly(3–hydroxybutyrate–co-3–hydroxyvalerate (PHBV) Actives concentration: 4–20 mg·mL⁻¹ Solution flow rate: 1 mL·min⁻¹ CO_2 flow rate: 1.2 L·h⁻¹	Particle size: 0.39–0.57 µm Morphology: spherical Structure: amorphous Polymorphic nature: – Precipitation yield: 13.25–49.71% Coprecipitation efficiency: 58.29–99.54%	
N–acetylcysteine	Micronization	Acetone/SC–CO_2	Pressure: 8–15 MPa Temperature: 308 K Drug concentration: 0.020–0.040 g·mL⁻¹ Solution flow rate: 1 mL·min⁻¹ CO_2 flow rate: 1.2 L·h⁻¹	Particle size: 2.9–301.9 µm Morphology: crystalline plate-like Structure: amorphous Polymorphic nature: – Precipitation yield: – Reduction in size: 57.48–99.59	
9-Nitro-camptothecin	Micronization	Dichloromethane Methanol and dimethyl formamide (DMF)/mixture, tetrahydrofuran (THF) SC–CO_2	Pressure: 8–12.5 MPa Temperature: 303–318 K Drug concentration: 0.5–1.1 g·L⁻¹ Solution flow rate: 0.3–1.2 mL·min⁻¹ CO_2 flow rate: 1.2 kg·h⁻¹	Particle size: 477–910 nm Morphology: spherical Structure: crystalline Polymorphic nature: + Precipitation yield: – Reduction in size: 99.55–99.77%	Huang et al. (2015)

(continued)

Table 2.1 (continued)

Substance	Process	Solvent/antisolvent	Operational parameters	Results	References
Astaxanthin	Coprecipitation	Acetone–Ethanol/SC–CO_2	Pressure: 8–15 MPa Temperature: 313–333 K Wall material: polyvinylpyrrolidone (PVP) Polymer concentration: 0.73–2.94 mg·mL^{-1} Drug concentration: 0.14 gL^{-1} Solution flow rate: – CO_2 flow rate: 0.9 L·h^{-1}	Particle size: 99.1–203.1 nm Morphology: Spherical Structure: amorphous Polymorphic nature: – Precipitation yield: – Encapsulation efficiency: –	Kaga et al. (2018)
Carbamazepine	Coprecipitation	Methanol/SC–CO_2 Ethanol/SC–CO_2 Dichloromethane/SC–CO_2 Dimethylsulfoxide/SC–CO_2	Pressure: 10–15 MPa Temperature: 313–333 K Drug concentration: 15–30 gL^{-1} Wall material: saccharine Solution flow rate: 1 mL·min^{-1} CO_2 flow rate: 1.2 kg·h^{-1}	Particle size: 5–10 μm Morphology: crystalline plate-like, needle-like Structure: – Polymorphic nature: I, I + II Precipitation yield: 45–65% Encapsulation efficiency: –	Cuadra et al. (2018)
Gefitinib	Micronization	Dichloromethane Ethanol/SC–CO_2	Pressure: 9–13 MPa Temperature: 308–323 K Drug concentration: 0.5–1 g·L^{-1} Solution flow rate: 0.5–2.5 mL·min^{-1} CO_2 flow rate: 1.2 kg·h^{-1}	Particle size: 0.6–2.51 μm Morphology: crystalline plate-like Structure: amorphous Polymorphic nature: I Precipitation yield: – Reduction in size: –	Liu et al. (2017)

(continued)

Table 2.1 (continued)

Substance	Process	Solvent/antisolvent	Operational parameters	Results	References
Gefitinib	Coprecipitation	Dichloromethane Ethanol/SC–CO_2	Pressure: MPa Temperature: K Drug concentration: 1 mg·mL^{-1} Wall material: poly (L-lactic acid) (PLLA) Solution flow rate: 1 mg·mL^{-1} CO_2 flow rate: 1.2 kg·h^{-1}	Particle size: 1.08–3.41 µm Morphology: – Structure: amorphous Precipitation yield: 0.92–12.32 Encapsulation efficiency: 10.8–87.78	Lin et al. (2017)
Insulin	Coprecipitation	Acetone-DMSO/SC–CO_2	Pressure: 10–20 MPa Temperature: 313 K Drug concentration: – Wall material: hydroxypropyl methyl cellulose phthalate (HPMCP) Solution flow rate: – CO_2 flow rate: 720 mL·h^{-1}	Particle size: 1–2 µm Morphology: flake Structure: amorphous Precipitation yield: 48.2–51.6% Encapsulation efficiency: 6.8–28.6%	Tandya et al. (2016)
Moxifloxacin	Micronization	Acetic acid/SC–CO_2 DMSO/SC–CO_2 N,N-dimethylformamide (DMFA)/SC–CO_2	Pressure: 15 MPa Temperature: 313 K Drug concentration: 1–50 mg·L^{-1} Solution flow rate: 1 mL·min^{-1} CO_2 flow rate: 3.0 kg·h^{-1}	Particle size: 0.6–1.2 µm Reduction in size: – Morphology: prismatic, needle-like crystals and flat polygonal sheets Structure: amorphous Polymorphic nature: – Precipitation yield: – Size reduction: 85–92.5	Kudryashova et al. (2017)

(continued)

Table 2.1 (continued)

Substance	Process	Solvent/antisolvent	Operational parameters	Results	References
Naproxen	Coprecipitation	Ethanol, Dichloromethane/SCCO$_2$	Pressure: 10–20 MPa Temperature: 313–323 K Wall material: Eudragit (EUD) and poly(l-lactic acid) (PLA) Drug concentration: 0.5–1.1 mg/mL Solution flow rate: 4 mL·min^{-1} CO$_2$ flow rate: 0.66 kg·h^{-1}	Particle size: 0.6–1.43 µm Reduction in size: – Morphology: spherical Structure: amorphous Precipitation yield: – Encapsulation efficiency: 2.95–13.24%	Montes et al. (2014)
Nimesulide	Coprecipitation	Dimethylsulfoxide/SC–CO$_2$	Pressure: 9–15 MPa Temperature: 308–313 K Drug concentration: 30 g·L^{-1} Wall material: polyvinylpyrrolidone (PVP) Solution flow rate: 1 mL·min^{-1} CO$_2$ flow rate: 1.8 kg·h^{-1}	Particle size: 1.67–4.70 µm Morphology: spherical Structure: amorphous Precipitation yield: 47.3–99.8% Encapsulation efficiency: %	Prosapio et al. (2016)
Tamoxifen	Coprecipitation	Dichloromethane/SC–CO$_2$	Pressure: 13 MPa Temperature: 311 K Wall material: poly-l-lactic acid (PLLA) Polymer concentration: 12.5 mg·mL^{-1} Drug concentration: 1 mg/mL^{-1} Solution flow rate: 1.3 mL·min^{-1} CO$_2$ flow rate: –	Particle size: – Morphology: Spherical Structure: amorphous Polymorphic nature: – Precipitation yield: – Encapsulation efficiency: – Reduction in size: –	

Table 2.2 Recent applications of SAF for enrichment of bioactive constituents focusing on the biorefining of plant matrices into marketable products

Bioactive compounds source	Bioactive compounds (%), w/w before SAF processing	Extraction solvent to obtain the bioactive compounds	SAF fractionation solvent	Operating parameters[a]
Terpenes from eucalyptus leaves extract	p-Cymene = 23.81 β-Pinene = 5.22 4 Terpineol = 4.78 Cryptone = 6.05 Piperitone = 4.47	$SC\text{-}CO_2$	Ethanol	$P = 10\text{-}20$ $T = 308\text{-}323$ $QBS = 0.12\text{-}0.3$ $QCO_2 = 0.6\text{-}1.2$
Curcuminoids from turmeric extracts	Curcuminoids = 1.78	Ethanol	Ethanol	$P = 10\text{-}20$; $T = 313\text{-}333$; QBS $= 0.03$; $QCO_2 = 0.5\text{-}0.8$
Quercetin from onion peels extract	Quercetin = 3.4–5.7	Ethanol	Ethanol	$P = 10\text{-}12$; $T = 313\text{-}333$ $QBS = 0.024$ $QCO_2 = 1.02$
Polyphenols from grape marc extract	Proanthocyanidins = 0.9–1.98 Gallic acid = 0.06 Catechin = 0.12 Epicatechin = 0.23 Quercetin = 0.11 Malvidin = 0.16 Procyanidin B2 = 0.18	Phosphate buffer at pH = 7.0	Ethanol	$P = 10\text{-}12$; $T = 313\text{-}318$ $QBS = 0.12$ $QCO_2 = -$
Polyphenols from grape waste extract	Gallic acid= Catechin= Epicatechin= Resveratrol=	Hexane	Ethanol	$P = 8\text{-}15$; $T = 308\text{-}333$; QBS $=$ $-QCO_2 = 2.38$

(continued)

Table 2.2 (continued)

Bioactive compounds source	Bioactive compounds (%), w/w before SAF processing	Extraction solvent to obtain the bioactive compounds	SAF fractionation solvent	Operating parameters[a]
Caffeine and catechins from tea leaves	Caffeine = 10–12.8 Epicatechin = 1.9–2.0 Epicatechin gallate = 3.3–4.0 Epigallocatechin = 2.6–5.8 Epigallocatechin gallate = 9.0–14.4	Ethyl lactate Ethanol	Ethyl lactate Ethanol	$P = 10$–30; $T = 323$–353 $VCO_2/VBS = 20$–40 L/L
Polyphenols from yarrow extracts	Total polyphenols = 0.053	Ethanol	Ethanol	$P = 10$–20; $T = 308$–323; QBS $= 0.06$–0.6; $QCO_2 = 1.2$
Polyphenols from mango leaves extract	Gallic acid = 0.00365 methyl gallate = 1.67×10^{-3} iriflophenone 3-C-β-D-glucoside = 0.124 iriflophenone 3-C-(2-O-p-hydroxybenzoyl)-β-D-glucoside = 0.049 Mangiferin = 0.054 iriflophenone-3-C-(2-O-galloryl)-β-D-glucoside = 1.99×10^{-3} quercetin 3-D-galactoside = 1.52×10^{-3} quercetin 3-β-D-glucoside = 6.82×10^{-3} quercetin-3-O-xyloside = 2.59×10^{-3} 1, 2, 3, 4, 6-penta-O-galloryl-β-D-glucose = 2.35×10^{-3} quercetin (aglycone) = 0.96×10^{-3}	Ethanol	Ethanol	$P = 10$–15; $T = 308$—323; QBS $= 0.03$–0.6; $QCO_2 = 0.6$–2.4
Polyphenols from mango by-products extracts	Mangiferin = 218.2×10^{-4} Isomangiferin = 3.6×10^{-4} Q-3-Ogalactoside = 9963.9×10^{-4} Q-3-O-glucoside = 5953.2×10^{-4} Q-3-O-xyloside = 1550.3×10^{-4} Q-3-O-arabinoside = 737.7×10^{-4} Quercetin = 480.8×10^{-4} Kaempferol = 77.2×10^{-4}	Aqueous acetone solution (80% v/v)	Ethanol	$P = 8$–15; $T = 308$–318 $QCO_2 = 1$; $QCO_2 = -$

(continued)

Table 2.2 (continued)

Bioactive compounds source	Products characteristic	Bioactive compounds (%), w/w after SAF processing	Purification yield increase (%), w/w[b]	Reference
Terpenes from eucalyptus leaves extract	S1 = powder with uniform particle distribution	p-Cymene = 30.66–67.78 β-Pinene = 2.48–7.31 4 Terpineol = 0.8–1.9 Cryptone = 5.68–16.01 Piperitone = 0.24–0.77	p-Cymene = 128.77–284.67 β-Pinene = 47.51–140.04 4 Terpineol = 16.74–39.75 Cryptone = 93.88–2.65 Piperitone = 5.54–17.23	Chinnarasu et al. (2015)
Curcuminoids from turmeric extracts	S1 = particles with some aggregation and flake-like porous structure	Curcuminoids = 31.6–55.8	Curcuminoids = 1775.28–3134.83	Osorio-Tobón et al. (2016)
Quercetin from onion peels extract	S1 = particles with some aggregation and flake-like porous structure	Quercetin = 12.9–26	Quercetin = 226.32–634.15	Zabot and Meireles (2016)
Polyphenols from grape marc extract	S1 = particles with aggregates	Proanthocyanidins = 1.47–6.91 Gallic acid = 0.24 Catechin = 0.43 Epicatechin = 1.2 Quercetin = 0.58 Malvidin = 0.91 Procyanidin B2 = 0.56	Proanthocyanidins = 53–627 Gallic acid = 70 Catechin = 58 Epicatechin = 88 Quercetin = 90 Malvidin = 94 Procyanidin B2 = 53	Natolino et al. (2016)
Polyphenols from grape waste extract	S1 = traces of bioactives collected in most of experiments S2 = relevance in bioactives	Gallic acid = 0.09–0.16 Catechin = 0.06–0.14 Epicatechin = 0.02–0.11 Resveratrol = 1.3–1.81	Gallic acid = –4–100 Catechin = –18–250 Epicatechin = 20–267 Resveratrol = 10–78	Marqués et al. (2013)
Caffeine and catechins from tea leaves	S1 = particles had irregular shapes, with dimensions in the order of 100 μm and a smooth surface	Caffeine = 0.82–4.40 Epicatechin = 2.13–2.54 Epicatechin gallate = 3.88–5.37 Epigallocatechin = 2.98–6.29 Epigallocatechin gallate = 10.7–16.3	Caffeine = 6.76–44.2 Epicatechin = 108–136 Epicatechin gallate = 116–151 Epigallocatechin = 113–109 Epigallocatechin gallate = 113–119	Villanueva-Bermejo et al. (2015)

(continued)

Table 2.2 (continued)

Bioactive compounds source	Products characteristic	Bioactive compounds (%), w/w after SAF processing	Purification yield increase (%), w/w[b]	Reference
Polyphenols from yarrow extracts	S1 = particles with irregular morphology S2 = kept at ambient pressure to recover the components soluble in the SC–CO2/organic solvent phase which do not precipitate in the precipitation vessel	Total polyphenols = 18.9–40.9 (precipitation vessel) Total polyphenols = 21–33.3 (S1)	Total polyphenols = 232.64–286.87 (precipitation vessel) Total polyphenols = 47.35–64.53 (S1)	Villanueva-Bermejo et al. (2017)
Polyphenols from mango leaves extract	Non identified	Gallic acid = 0.224–0.945 methyl gallate = 0.399–1.289 iriflophenone 3–C–β–D–glucoside = 16.05–46.31 iriflophenone 3–C–(2–O–p–hydroxybenzoyl)–β–D–glucoside = 3.50–46.32 Mangiferin = 5.01–20.69 iriflophenone-3–C–(2–O–galloryl)–β–D–glucoside = 2.61–8.74 quercetin 3–D–galactoside = 0.58–2.39 quercetin 3–β–D–glucoside = 0.712–2.99 quercetin–3–O–xyloside = 0.558–2.36 1, 2, 3, 4, 6–penta–O–galloryl–β–D–glucose = 0.069–3.25 quercetin (aglycone) = 0.574–2.71	Gallic acid = 61.37–245.60 methyl gallate = 238.92–771.86 iriflophenone 3–C–β–D–glucoside = 129.51–373.72 iriflophenone 3–C–(2–O–p–hydroxybenzoyl)–β–D–glucoside = 71.21–408.03 Mangiferin = 91.12–376.26 iriflophenone–3–C–(2–O–galloryl)–β–D–glucoside = 130.99–438.06 quercetin 3–D–galactoside = 379.61–1532.89 quercetin 3–β–D–glucoside = 104.40–439.3 quercetin–3–O–xyloside = 215.44–911.97 1, 2, 3, 4, 6–penta–O–galloryl–β–D–glucose = 29.36–1385.11 quercetin (aglycone) = 597.92–2821.88	Guamán-Balcázar et al. (2017)

(continued)

Table 2.2 (continued)

Bioactive compounds source	Products characteristic	Bioactive compounds (%), w/w after SAF processing	Purification yield increase (%), w/w[b]	Reference
Polyphenols from mango by-products extracts	S1 = obtained of spherical particles S2 = used to recover liquid solvent	Mangiferin = 0.278–0.298 Isomangiferin = 3.6–4.8 × 10–3 Q-3–Ogalactoside = 10.555–11.492 Q-3–O-glucoside = 7.011–7.706 Q-3–O-xyloside = 1.742–1.917 Q-3–O-arabinoside = 0.892–0.993 Quercetin = 0.503–0.612 Kaempferol = 0.1–0.112	Mangiferin = 127.64–136.76 Isomangiferin = 119.44–133.33 Q-3–Ogalactoside = 105.94–115.35 Q-3–O-glucoside = 117.78–129.45 Q-3–O-xyloside = 112.37–123.62 Q-3–O-arabinoside = 120.96–134.65 Quercetin = 104.56–127.29 Kaempferol = 130.57–144.43	Meneses et al., (2015)

[a]T = Temperature (K); P = Pressure (MPa); QCO$_2$ = Carbon dioxide mass flow rate (kg·h^{-1}); VBS = Volume of bioactive compounds solution (mL); QBS = Bioactive compounds solution flow rate (mL·min^{-1}); S1 = Precipitation chamber (Catchpole et al. 2004) or precipitator vessel (Floris et al. 2010); S2 = cylindrical separator (Catchpole et al. 2004; Floris et al. 2010) and S3 = Term related to the 1st, 2nd and 3rd cylindrical separators, if available

[b]Purification yield increase (%), w/w = [Bioactive compounds (%), w/w after SAF processing—Bioactive compounds (%), w/w before SAF processing/Bioactive compounds (%), w/w before SAF processing

References

Adami R, Di Capua A, Reverchon E (2017) Supercritical assisted atomization for the production of curcumin-biopolymer microspheres. Powder Technol 305:455–461

Albarelli JQ, Santos DT, Cocero MJ, Meireles MAA (2016) Economic analysis of an integrated annatto seeds-sugarcane biorefinery using supercritical CO_2 extraction as a first step. Materials 6:494

Almeida RA, Rezende RVP, Cabral VF, Noriler D, Meier HF, Cardozo-Filho L, Cardoso FAR (2016) The effect of system temperature and pressure on the fluid-dynamic behavior of the supercritical antisolvent micronization process: a numerical approach. Braz J Chem Eng 33:73–90

Anwar M, Ahmad I, Warsi MH, Mohapatra S, Ahmad N, Akhter S, Ali A, Ahmad FJ (2015) Experimental investigation and oral bioavailability enhancement of nano-sized curcumin by using supercritical anti-solvent process. Eur J Pharm Biopharm 96:162–172

Arango-Ruiz Á, Martin Á, Cosero MJ, Jiménez C, Londoño J (2018) Encapsulation of curcumin using supercritical antisolvent (SAS) technology to improve its stability and solubility in water. Food Chem 258:156–163

Benelli P, Rosso Comim SR, Vladimir Oliveira J, Pedrosa RC, Ferreira SRS (2014) Phase equilibrium data of guaçatonga (Casearia sylvestris) extract + ethanol + CO_2 system and encapsulation using a supercritical anti-solvent process. J Supercrit Fluids 93:103–111

Catchpole OJ, Grey J, Mitchell K, Lan J (2004) Supercritical antisolvent fractionation of propolis tincture. J Supercrit Fluids 29(1–2):97–106

Chen C-R, Lin D-M, Chang C-MJ, Chou H-N, Wu J-J (2017) Supercritical carbon dioxide anti-solvent crystallization of fucoxanthin chromatographically purified from Hincksia mitchellae P.C. Silva. J Supercrit Fluids 119:1–8

Cheng Y-S, Lu P-M, Huang C-Y, Wu J-J (2017) Encapsulation of lycopene with lecithin and α-tocopherol by supercritical antisolvent process for stability enhancement. J Supercrit Fluids 130:246–252

Chinnarasu C, Montes A, Fernandez-Ponce MT, Casas L, Mantell C, Pereyra C, de la Ossa EJM, Pattabhi S (2015) Natural antioxidant fine particles recovery from Eucalyptus globulus leaves using supercritical carbon dioxide assisted processes. J Supercrit Fluids 101:161–169

Cuadra IA, Cabañas A, Cheda JAR, Pando C (2018) Polymorphism in the co-crystallization of the anticonvulsant drug carbamazepine and saccharin using supercritical CO_2 as an anti-solvent. J Supercrit Fluids 136:60–69

Dal Magro C, Aguiar GPS, Veneral JG, dos Santos AE, de Chaves LMPC, Oliveira JV, Lanza M (2017) Co-precipitation of trans-resveratrol in PHBV using solution enhanced dispersion by supercritical fluids technique. J Supercrit Fluids 127:182–190

Floris T, Filippino G, Scrugli S, Pinna MB, Argiolas F, Argiolas A, Murru M, Reverchon E (2010) Antioxidant compounds recovery from grape residues by a supercritical antisolvent assisted process. J Supercrit Fluids 54(2):165–170

Guamán-Balcázar MC, Montes A, Pereyra C, de la Ossa EM (2017) Precipitation of mango leaves antioxidants by supercritical antisolvent process. J Supercrit Fluids 128:218–226

Huang Y, Wang H, Liu G, Jiang Y (2015) New polymorphs of 9-nitro-camptothecin prepared using a supercritical anti-solvent process. Int J Pharm 496(2):551–560

Kaga K, Honda M, Adachi T, Honjo M, Wahyudiono, Kanda H, Goto M (2018) Nanoparticle formation of PVP/astaxanthin inclusion complex by solution-enhanced dispersion by supercritical fluids (SEDS): effect of PVP and astaxanthin Z-isomer content. J Supercrit Fluids 136:44–51

Kudryashova EV, Sukhoverkov KV, Deygen IM, Vorobei AM, Pokrovskiy OI, Parenago OO, Presnov DE, Egorov AM (2017) Moxifloxacin micronization via supercritical antisolvent precipitation. Russ J Phys Chem B 11(7):1153–1162

Kurniawansyah F, Mammucari R, Foster NR (2017) Polymorphism of curcumin from dense gas antisolvent precipitation. Powder Technol 305:748–756

Lin K-L, Chng L-M, Lin JC-T, Hsu S-L, Young C-C, Shieh C-J, Chang C-MJ (2014) Effect of anti-solvent conditions on low density supercritical fluids precipitation of zeaxanthin palmitates from lycium barbarum fruits. J Supercrit Fluids 87:104–110. https://doi.org/10.1016/j.supflu.2014.01.001

Lin Q, Liu G, Zhao Z, Wei D, Pang J, Jiang Y (2017) Design of gefitinib-loaded poly (l-lactic acid) microspheres via a supercritical anti-solvent process for dry powder inhalation. Int J Pharm 532(1):573–580

Liu G, Lin Q, Huang Y, Guan G, Jiang Y (2017) Tailoring the particle microstructures of gefitinib by supercritical CO_2 anti-solvent process. J CO_2 Util 20: 43–51

Machado APDF, Rueda M, Barbero GF, Martín Á, Cocero MJ, Martínez J (2018) Co-precipitation of anthocyanins of the extract obtained from blackberry residues by pressurized antisolvent process. J Supercrit Fluids 137:81–92

Marqués JL, Porta GD, Reverchon E, Renuncio JAR, Mainar AM (2013) Supercritical antisolvent extraction of antioxidants from grape seeds after vinification. J Supercrit Fluids 82:238–243

Meneses MA, Caputo G, Scognamiglio M, Reverchon E, Adami R (2015) Antioxidant phenolic compounds recovery from Mangifera indica L. by-products by supercritical antisolvent extraction. J Food Eng 163:45–53

Miao H, Chen Z, Xu W, Wang W, Song Y, Wang Z (2018) Preparation and characterization of naringenin microparticles via a supercritical anti-Solvent process. J Supercrit Fluids 131:19–25

Montes A, Kin N, Gordillo MD, Pereyra C, de la Ossa EJM (2014) Polymer–naproxen precipitation by supercritical antisolvent (SAS) process. J Supercrit Fluids 89:58–67

Montes A, Wehner L, Pereyra C, Martínez de la Ossa EJ (2016a) Generation of microparticles of ellagic acid by supercritical antisolvent process. J Supercrit Fluids 116:101–110

Montes A, Wehner L, Pereyra C, Martínez de la Ossa EJ (2016b) Precipitation of submicron particles of rutin using supercritical antisolvent process. J Supercrit Fluids 118:1–10

Montes A, Wehner L, Pereyra C, de la Ossa EJM (2016c) Mangiferin nanoparticles precipitation by supercritical antisolvent process. J Supercrit Fluids 112:44–50

Natolino A, Da Porto C, Rodríguez-Rojo S, Moreno T, Cocero MJ (2016) Supercritical antisolvent precipitation of polyphenols from grape marc extract. J Supercrit Fluids 118:54–63

Osorio-Tobón JF, Carvalho PIN, Rostagno MA, Petenate AJ, Meireles MAA (2016) Precipitation of curcuminoids from an Ethanolic turmeric extract using a supercritical antisolvent process. J Supercrit Fluids 108:26–34

Prosapio V, Reverchon E, De Marco I (2016) Formation of PVP/nimesulide microspheres by super-critical antisolvent coprecipitation. J Supercrit Fluids 118:19–26. https://doi.org/10.1016/j.supflu.2016.07.023

Santana ÁL, Santos DT, Meireles MAA (2019) Perspectives on small-scale integrated biorefineries using supercritical CO_2 as a green solvent. Curr Opin Green Sustain Chem 18:1–12

Tandya A, Zhuang HQ, Mammucari R, Foster NR (2016) Supercritical fluid micronization techniques for gastro-resistant insulin formulations. J Supercrit Fluids 107:9–16

Torres RAC, Santana ÁL, Santos DT, Meireles MAA (2016) Perspectives on the application of supercritical antisolvent fractionation process for the purification of plant extracts: effects of operating parameters and patent survey. Recent Patents Eng 10:121–130

Villanueva-Bermejo D, Ibáñez E, Reglero G, Turner C, Fornari T, Rodriguez-Meizoso I (2015) High catechins/low caffeine powder from green tea leaves by pressurized liquid extraction and supercritical antisolvent precipitation. Sep Purif Technol 148:49–56

Villanueva-Bermejo D, Zahran F, Troconis D, Villalva M, Reglero G, Fornari T (2017) Selective precipitation of phenolic compounds from Achillea millefolium L. extracts by supercritical anti-solvent technique. J Supercrit Fluids 120:52–58

Zabihi F, Xin N, Li S, Jia J, Cheng T, Zhao Y (2014) Polymeric coating of fluidizing nano-curcumin via anti-solvent supercritical method for sustained release. J Supercrit Fluids 89:99–105

Zabot GL, Meireles MAA (2016) On-line process for pressurized ethanol extraction of onion peels extract and particle formation using supercritical antisolvent. J Supercrit Fluids 110:230–239

Chapter 3
Integrated Biorefinery Approach for the Valorization of Plant Materials Using Supercritical Antisolvent-Based Precipitation Technique for Obtaining Bioactive Compounds

Abstract This work investigates a novel approach for turmeric rhizomes valorization for the obtaining of microparticles composed of curcuminoids ethanolic extract and fractionated volatile oils, using compressed carbon dioxide as an antisolvent and the recovery of phenolic compounds and carbohydrates with pressurized hot water. In addition, a cheap and versatile method for the quantification curcuminoids using thin-layer chromatography coupled to image processing analysis was applied to the solid wastes and liquid extracts from turmeric, derived from extraction processes which employed supercritical CO_2 and pressurized liquid ethanol. Coprecipitation of PEG with compressed CO_2 resulted in the formation of spherical particles and blocks of aggregates. The reaction with vanillin-sulfuric acid reagent favored the detection of curcuminoids with a reasonable sensibility and wide linear range and mean recoveries between 92 and 104%.

3.1 Introduction

Researches on the use of natural substances to replace the synthetic ones emerged for the valorization of food matrices. Turmeric (*Curcuma longa L.*) is composed of health-beneficial substances, such as glucose, starch, phenolic and volatile compounds, and unsaturated fatty acids (Rosso and Mercadante 2009; Santana et al. 2017a).

Environmentally friendly processes coupled with green solvents (carbon dioxide, ethanol, and water) are relatively new and promising techniques to obtain valuable biological substances from different botanical sources (Xu et al. 2016). These procedures were successfully used to modify polymeric fraction of waste annatto seeds (Alcázar-Alay et al. 2017), to recover phenolic compounds and sugars from turmeric wastes (Santana and Meireles 2016; Santana et al. 2017c), to obtain energy from paper sludge powder (Xu and Lancaster 2008), and isolate of phenolic compounds from black carrot (Aşkin Uzel 2017).

The obtaining of particles from plant extracts using coprecipitation procedures with carbon dioxide was successfully applied for the obtained products with improved composition and stability (Santos and Meireles 2013).

Thin-layer chromatography is one of the cheapest, quickest and most efficient separation methods for many classes of chemical compounds. It has some important advantages over other chromatographic techniques like low cost of instrumentation, evaluation of the whole sample because of the spatial separation, the ability of making simultaneous separations (even 20 samples) and, of course, shortened time required for analysis. All these make thin-layer chromatography a convenient choice for many applications including analytical, biomedical, and pharmaceutical field (Soponar et al. 2008).

In this context, this work evaluates a novel approach for the valorization of turmeric rhizomes to obtain microparticles composed of curcuminoids and volatile oils, using compressed carbon dioxide as an antisolvent, and to recover phenolic compounds and carbohydrates by partial hydrolysis with pressurized hot water. In addition, a cheap and versatile method to detect phenolic compounds using thin-layer chromatography coupled to image processing analysis was applied to the solid wastes and liquid extracts from turmeric, derived from extraction processes which employed supercritical CO_2 and pressurized liquid ethanol.

3.2   Materials and Methods

3.2.1   Materials

Crudeturmeric (CT) was purchased from the Oficina de ErvasFarmácia de Manipulação Ltda (lot 065DM, RibeirãoPreto, Brazil).

3.2.2   Supercritical Fluid and Pressurized Liquid Extraction

The CT was extracted with supercritical CO_2 at 60 °C and 25 MPa (Carvalho et al. 2014), resulting on two types of extracts: light-phase turmeric volatile oil (LTO), a visually homogenous extract, and heavy-phase turmeric volatile oil (HTO), a viscous extract with the presence of solid and aqueous fractions.

From deflavored turmeric (DT), curcuminoids were extracted with pressurized ethanol liquid ethanol at 60 °C and 10 MPa(Osorio-Tobón et al. 2014) producing a curcuminoids ethanolic extract.

The resulted deflavored and depigmented turmeric (DDT) was subjected to partial hydrolysis using pressurized hot water extraction, from which conditions evaluated in this work were 1 and 7 MPa and 40, 70 and 100 °C (Santana et al. 2017b). Besides these procedures, the starches from the solid fraction obtained at each process stage were recovered as an alternative to reuse these waste materials, according to (Santana et al. 2017c). The starch from the waste subjected to hot water extraction was not recovered because of the poor availability of samples.

3.2.3 Coprecipitation of Turmeric Extracts

Microparticles of fractionated turmeric volatile oils and curcuminoids extracts were obtained through coprecipitation at 20 °C using polyethelene glycol (PEG, 10,000 g/mol, Sigma-Aldrich, Darmstadt, Germany) as wall material, dichloromethane (Synth, Diadema, Brazil) as a solvent and compressed CO_2 as antisolvent (White Martins, Campinas, Brazil). The proportions of extract and wall material evaluated were 1/10 (w/w) and 5/10 (w/w) (Santana and Meireles 2017).

3.2.4 Scanning Electron Microscopy (SEM)

The structure of turmeric products was examined using a scanning electron microscope (Leo 440i, Cambridge, England), accelerating potential of 15 kV, current of 50 pA, and resolution of 1500×. The samples were applied on circular aluminum stubs with double carbon sticky tape and coated with 200 Å of gold on the Sputter Coater (EMITECH, K450, Kent, United Kingdom).

3.2.5 Thin-Layer Chromatography

3.2.5.1 Chemicals and Standard Solutions

Standards of curcumin (≥94% curcuminoids; ≥80% curcumin), demethoxycurcumin, bisdemethoxycurcumin were purchased from Sigma-Aldrich (Darmstadt, Germany). The curcuminoids standards were diluted in ethanol (99.5%, Chemco, Hortolandia, Brazil) until the concentration of 10 mg/mL. Turmeric products were diluted until 30 mg/mL

The mobile phase was composed of chloroform (Merck, Darmstadt, Germany), ethanol, and glacial acetic acid (Synth, Diadema, Brazil), using the proportion of 95/05/01(v/v/v).

The vanillin-sulfuric acid spray reagent was prepared, according to the formulation proposed by Krishnaswamy (2003) by dilution of 0.5 g of vanillin (Sigma-Aldrich, Darmstadt, Germany) that was diluted in 20 ml of ethanol and 80 ml of sulfuric acid (Exodo, Hortolandia, Brazil).

3.2.5.2 Thin-Layer Chromatography with Image Background

This method starts with image acquisition with subsequent processing. The chromatographic plates were recorded using a 5 megapixel still camera, using distance of 30 cm, and adequate illumination. The software ImageJ (Ferreira and Rasband

2010) was used for digital processing of images of captures images and quantitative evaluation of gray intensity value obtained by chromatographic spots inserted on TLC plates, according to the procedures performed elsewhere (Johner and Meireles 2016).

3.2.5.3 Method Validation

For the calibration, solutions containing the standards curcumin, demethoxycurcumin, and bisdemethoxycurcumin (10 mg/mL) were applied in duplicate as spots in increasing volumes. The calibration functions were estimated for each of the investigated compounds by plotting the quantity of standard versus the calculated peak area values, expressed in pixels. The areas of the peaks, expressed in pixels, were calculated in order to estimate the concentrations of curcuminoids present in each product.

The linearity range was characterized by the linear regression equation, the correlation coefficient (R^2) of determination, and standard deviation (s). The calculations of the limit of detection (LDQ) and limit of quantification (LOQ) were performed according to the procedure proposed elsewhere(Shrivastava and Gupta 2011). The intraday precision was measured in two intervals after the acquisition of images (30 and 60 min). The chosen period was stated because of low stability of the compounds in the silica gel plates, observed in preliminary assays. The interday precision was measured considering two consecutive days.

The validation and accuracy of the method were investigated on by establishing comparisons with turmeric products, from which curcuminoids were quantified elsewhere using high-performance liquid chromatography (HPLC), i.e., turmeric extracts fractionated volatile oils (Santana et al. 2017a), curcuminoids ethanolic extract and solid products derived from extraction processes that employed supercritical carbon dioxide and pressurized liquid ethanol (Santana et al. 2017c).

The HPLC method used to quantify curcuminoids established elsewhere (Osorio-Tobón et al. 2016) was applied using a flow rate of 1.25 mL/min., a Waters Alliance separation module (269SD, Milford, USA), a diode array detector (2998), and a C18 column (150 × 4.6 nm, id., 2.6 μm, Phenomenex, Torrance, USA) that was maintained at 50 °C. The samples were diluted in ethanol at 1 mg/mL. The mobile phases were solvent A, which consisted of 0.1% glacial acetic acid (Ecibra, Brazil) in Milli-Q water (Millipore®), and solvent B, which consisted of 0.1% glacial acetic acid in acetonitrile (JT Baker, USA). The curcuminoids were separated by increasing the content of solvent B from 45 to 65% at a constant flow rate.

Established the validation of this TLC couples to image background estimation, this method was applied to quantify curcuminoids from turmeric coprecipitates, which were not quantified previously.

3.3 Results and Discussion

3.3.1 Scanning Electron Microscopy Analysis

Coprecipitation of PEG with compressed carbon dioxide resulted in the formation of spherical particles, and blocks of aggregates. Besides this, SEM images show that high proportion of wall material led to decreased agglomerated particles, as can be observed in Fig. 3.1), similar to that obtained to annatto extract coprecipitates (Santos and Meireles 2013).

The high agglomeration with partial loss of spherical shape from the microparticles obtained at a mass ratio of 5/10 (w/w) occurred probably because the amount of polymer was not sufficient to effectively cover the amount of extract (Santana and Meireles 2017).

Starch granule from crude turmeric is provided by a rigid structure with the presence of cellulosic walls (Fig. 3.2). Supercritical CO_2 disrupted the cellulosic walls, contributing to the increasing of availability granule for subsequent extraction procedures. Pressurized liquid ethanol increases the exposition of the granules, and contributed for their dispersing (Fig. 3.2).

The SEM photos of recovered starches suggest that part of the granules previously available was left behind because of successive washings performed in the isolation of starch (Santana et al. 2017c).

Partial hydrolysis with pressurized hot water promoted the pre-gelatinization of starch granules. The conditions employed enhanced the diffusion of water into the granules. Nevertheless, the conditions of 100 °C and 7 MPa induced the leaching of soluble portion of starch into water, transforming the granules into formless sacs (Fig. 3.3).

3.4 Thin-Layer Chromatography Coupled to Image Estimation Background

3.4.1 Method Validation

A number of software packages for image analysis that are suitable for the evaluation of bioactive constituents from diverse products, like triterpene acids (Świeboda et al. 2014) and bixin (Johner and Meireles 2016) from plant extracts and glycerol from biodiesel (Bansal et al. 2008).

Concentration profiles of selected lanes can also be displayed and analyzed by these programs and quantification can be performed either by a two-dimensional method (i.e., by computing the distance, and mean gray value of a selected spot or band on the image) or by a one-dimensional approach (peaks present in concentration profiles are subjected to quantification). Besides these, chromatographic properties,

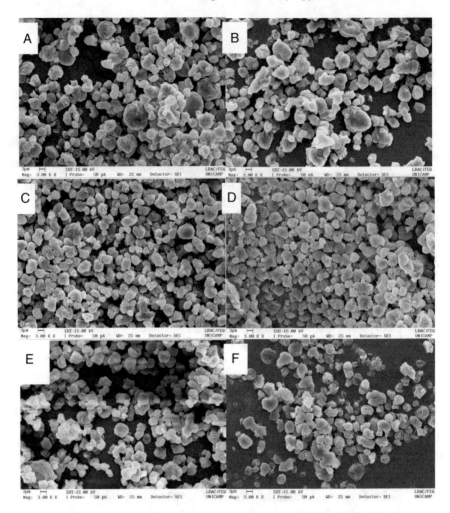

Fig. 3.1 SEM images of turmeric extracts coprecipitates in PEG: curcuminoids ethanolic extract (1extract/10 PEG, w/w, **a** and 5extract/10 PEG, w/w–**b**) and fractionated turmeric volatile oils LTO (1extract/10 PEG, w/w, **c** and 5extract/10 PEG, w/w –**d** and HTO (1extract/10 PEG, w/w, **e** and 5extract/10 PEG, w/w)

such as retardation factor (R_F), spot area and resolution can also be calculated (Sherma and Fried 2003).

The reaction with vanillin-sulfuric acid reagent allowed the detection of curcuminoids (red bands) and three peaks, which intensity was expressed in terms of grayscale and distance (Figs. 3.4 and 3.5). The coprecipitated extracts were not evaluated because spots were not detected in visible zone.

A solvent of the correct strength for a single development separation will migrate the sample into the R_F range 0.2–0.8, or thereabouts, and if of the correct selectivity,

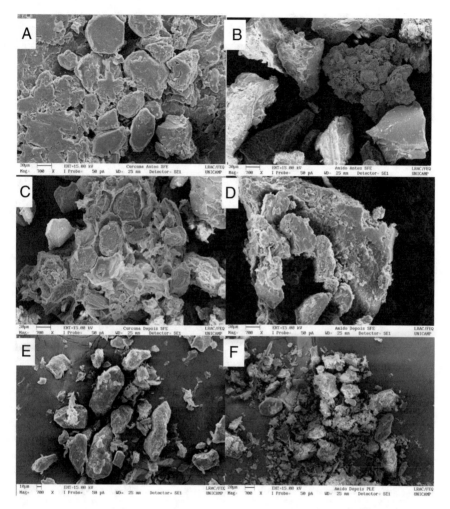

Fig. 3.2 SEM images of CT (**a**), CT starch (**b**), DT (**c**), DT starch (**d**), DDT (**e**) and DDT starch (**f**)

will distribute the sample components evenly throughout this range. The resulted R_Fs denote the adequate separation of the compounds (Table 3.1).

The calibration equation for each of the investigated compounds was determined by plotting the obtained peak area versus concentration using four increasing volumes of standard solutions. The calibration functions were linear in the concentration range from 0.15 to 0.30 mg/spot for curcuminoids.

The good linearity in the corresponding concentration range was evaluated by the linear regression equations and the values of the correlation coefficient (R^2), standard deviation (s), limit of detection (LDQ) and limit of quantification (LOQ) presented in Table 3.2.

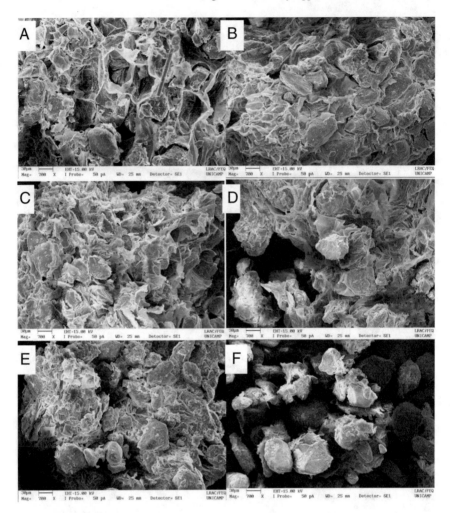

Fig. 3.3 SEM images turmeric products obtained from partial hydrolysis DDT (a), 1 MPa and 40 °C (b), 7 MPa and 40 °C (c), 1 MPa and 70 °C (d), 7 MPa and 70 °C (e),1 MPa and 100 °C (f), 7 MPa and 100 °C (g)

Limit of detection (LDQ) and the limit of quantification (LOQ) of curcuminoids were in the range of 0.15–0.30 mg/spot (Table 3.2).

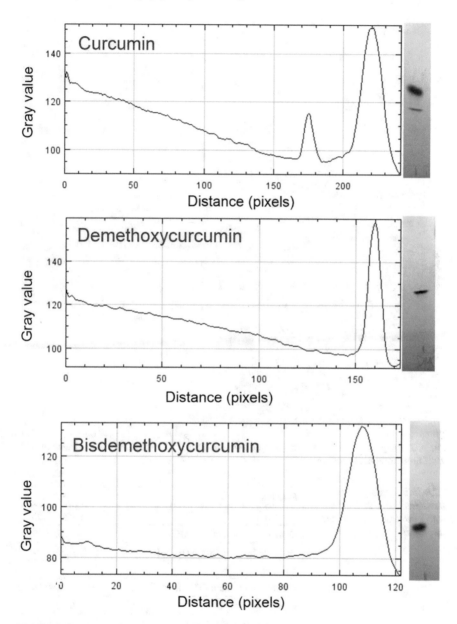

Fig. 3.4 Integrated peak area from curcuminoids standards on TLC plates after spot detection with vanillin-sulfuric acid spray reagent

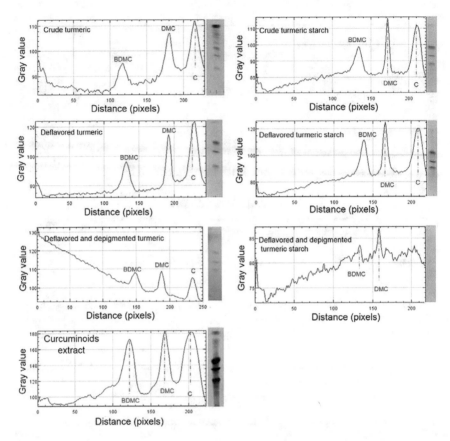

Fig. 3.5 Integrated peak area from turmeric products on TLC plates after spot detection with vanillin-sulfuric acid spray reagent

Table 3.1 Identity of the standards in the TLC plates with 8 cm elution height and R_F values

Standards	R_F (−)
Curcumin (C)	0.44–0.55
Demethoxycurcumin (DMC)	0.42
Bisdemethoxycurcumin (BDMC)	0.33
Products	
Curcuminoidsethanolic extract (CEE)	0.41–0.64
Crude turmeric (CT)	0.32–0.93
Deflavoredturmeric (DT)	0.39–0.93
Deflavored and depigmented turmeric (DDT)	0.43–0.68
CT starch	0.39–0.90
DT starch	0.39–0.89
DDTstarch	–

Table 3.2 Method evaluation parameters (linearity range, regression equation, correlation coefficient (R^2), standard error of estimate (s), limit of detection (LDQ) and limit of quantification (LOQ)

Compound	Linearity range (mg/spot)	Regression equation	R^2	s	LDQ (mg/spot)	LOQ (mg/spot)
C	0.15–0.30	$y = 2 \times 10^{-6} x - 6 \times 10^{-4a}$	0.99	0.01	0.04	0.15
DMC	0.15–0.30	$y = 2 \times 10^{-6} x - 7 \times 10^{-4a}$	0.99	0.01	0.04	0.12
BDMC	0.15–0.30	$y = 3 \times 10^{-6} x - 7 \times 10^{-4a}$	0.99	1×10^{-3}	0.02	0.06

[a] Where y is the amount of compound (mg) and x is the intensity measured in pixels

3.4.2 Precision and Accuracy

The intermediate precision of the method expressed as percent relative standard deviation (RSD %) was determined at four concentration levels for the curcuminoids standards (Table 3.3). The results of the intermediate precision of the assay (Table 3.3) show that the method has good precision, with RSDs lower than 10%, according to those obtained for the determination of food dyes (Soponar et al. 2008) and acidic catecholamine metabolites (Casoni et al. 2014) using TLC coupled to image background estimation.

The accuracy of the method, expressed by percent of mean recovery rates of each compound was determined in the range between 92.29 and 104.17%, which values were comparable to those obtained to quantify capsaicinoids in chili samples (Dawan et al. 2017).

3.4.3 Application of the Method to Turmeric Products

The applicability of the proposed TLC coupled to image estimation background to determine the level of curcuminoids in turmeric products was also investigated. The chromatographic separation, detection with vanillin-sulfuric acid spray reagent, and image processing analysis were performed under the same conditions applied for the standards (Table 3.4).

The amount of each compound calculated by the linear regression equation present in turmeric products as compared to the amount of each standard, and the recovery rate of the method was calculated. The results show recovery rates between 30.97 and 119.93%.

Table 3.3 Accurancy of the TLC with image background estimation

	Bisdemethoxycurcumin		
Addedamount (mg/spot)	Intradayprecision (RSD%)	Interdayprecision (RSD%)	Mean recovery (%)
0.30	5.15	5.13	98.05
0.25	5.17	5.15	92.62
0.15	5.19	5.16	93.84
0.10	5.22	5.20	92.29
Curcumin			
Addedamount (mg/spot)	Intradayprecision (RSD%)	Interdayprecision (RSD%)	Mean recovery (%)
0.30	8.04	8.34	101.19
0.25	8.04	8.33	101.15
0.15	8.04	8.29	100.99
0.10	8.05	8.23	100.79
Demethoxycurcumin			
Addedamount (mg/spot)	Intradayprecision (RSD%)	Interdayprecision (RSD%)	Mean recovery (%)
0.30	8.02	8.95	104.17
0.25	8.03	8.67	104.09
0.15	8.01	8.92	103.95
0.10	8.00	8.90	103.74

Table 3.4 Recovering of curcuminoids from turmeric products

Product	BDMC (%)	C (%)	DMC (%)
Curcuminoidsextract	54.65	77.11	57.89
Crude turmeric (CT)	77.11	101.78	108.72
Deflavoredturmeric (DT)	115.65	111.00	119.93
Deflavored and depigmented turmeric (DDT)	37.95	61.13	36.28
CT starch	44.54	41.83	41.12
DT starch	54.03	45.72	47.08
DDTstarch	35.16	34.96	30.97

Quantification of bioactives using thin-layer chromatography (TLC) coupled image analysis is attractive for its low cost and convenience. The non-uniformity of TLC plate Illumination during the acquisition of photos probably promoted discrete deviation (Zhang 2006) from the data obtained with HPLC method.

Despite this, the results obtained using the alternative method were in good agreement with those obtained by HPLC method, with standard deviation values (*s*) lower than 10% (Table 3.5).

Table 3.5 Comparison of TLC-IBG method used in this to the HPLC applied for curcuminoids quantification

	Total (g/100 g)			BDMC (%)			C (%)			DMC (%)		
	HPLC	TLC/IBE	s	HPLC	TLC/IBE	s	HPLC	TLC/IBE	s	HPLC	TLC/IBE	s
CT	11.33	10.17	0.82	39.08	37.66	1.01	43.11	41.58	1.08	17.81	20.77	2.09
DT	12.45	1.19	1.60	35.12	34.43	0.49	46.07	43.94	1.51	18.81	21.63	1.99
DDT	4.12	4.08	0.03	43.50	41.54	1.39	31.09	32.94	1.31	25.41	25.53	0.08
CT starch	6.72	3.75	2.10	23.15	25.63	1.76	53.91	52.03	1.33	22.94	22.34	0.42
DT starch	9.96	4.31	4.00	44.70	39.26	3.85	36.63	38.27	1.16	18.67	22.47	2.69
DDT starch	0.30	0.30	0.00	36.86	39.07	1.56	48.47	45.55	2.06	14.67	15.38	0.50
Curcuminoidsextract	12.08	11.83	0.18	37.53	40.36	2.00	42.77	40.16	1.85	19.70	19.48	0.15

3.5 Conclusions

Integrated processes for the coprecipitation of extracts and the use of starch fraction from turmeric wastes were performed on the possibility to valorize turmeric by providing value-added products and support researches on the complete use of this raw material. Coprecipitation of turmeric extracts with polyethylene glycol by compressed carbon dioxide resulted in the formation of spherical particles, and blocks of aggregates. The reaction with vanillin-sulfuric acid reagent favored the detection of curcuminoids with a reasonable sensibility and wide linear range and mean recoveries between 92 and 104%.

Acknowledgements Diego T. Santos thanks CNPq (processes 401109/2017-8; 150745/2017-6) for the post-doctoral fellowship. Ádina L. Santana thanks CAPES (1764130) for the post-doctoral fellowship. M. Angela A. Meireles thanks CNPq for the productivity grant (302423/2015-0). The authors acknowledge the financial support from CNPq (process 486780/2012-0) and FAPESP (processes 2012/10685-8; 2015/13299-0).

References

Alcázar-Alay SC, Osorio-Tobón JF, Forster-Carneiro T, Steel CJ, Meireles MAA (2017) Polymer modification from semi-defatted annatto seeds using hot pressurized water and supercritical CO_2. J Supercrit Fluids 129:48–55

Aşkin Uzel R (2017) A practical method for isolation of phenolic compounds from black carrot utilizing pressurized water extraction with in-site particle generation in hot air assistance. J Supercrit Fluids 120 2:320–327

Bansal K, McCrady J, Hansen A, Bhalerao K (2008) Thin layer chromatography and image analysis to detect glycerol in biodiesel. Fuel 87:3369–3372

Carvalho PIN, Osório-Tobón JF, Rostagno MA, Petenate AJ, Meireles MAA (2014) Optimization of the ar-turmerone extraction from turmeric (*Curcuma longa L.*) using supercritical carbon dioxide. 14th European meeting on supercritical fluids, Marseille—France

Casoni D, Sima IA, Sârbu C (2014) Thin-layer chromatography–an image-processing method for the determination of acidic catecholamine metabolites. J Sep Sci 37:2675–2981

Dawan P, Satarpai T, Tuchinda P, Shiowatana J, Siripinyanond A (2017) A simple analytical platform based on thin-layer chromatography coupled with paper-based analytical device for determination of total capsaicinoids in chilli samples. Talanta 162:460–465

Ferreira T, Rasband W (2010) The ImageJ user guide. IJ 1.45

Johner JCF, Meireles MAA (2016) Construction of a supercritical fluid extraction (SFE) equipment: validation using annatto and fennel and extract analysis by thin layer chromatography coupled to image. Food Sci Tech (Campinas) 36:1–38

Krishnaswamy NR (2003) Chemistry of natural products: a laboratory handbook. Universities Press, Hydebarat

Osorio-Tobón JF, Carvalho PIN, Barbero GF, Nogueira GC, Rostagno MA, Meireles MADA (2016) Fast analysis of curcuminoids from turmeric (*Curcuma longa L.*) by high-performance liquid chromatography using a fused-core column. Food Chem 200:167–174

Osorio-Tobón JF, Carvalho PIN, Rostagno MA, Petenate AJ, Meireles MAA (2014) Extraction of curcuminoids from deflavored turmeric (*Curcuma longa L.*) using pressurized liquids: process integration and economic evaluation. J Supercrit Fluids 95:167–174

Rosso VV, Mercadante AZ (2009) Dyes in South America. In: Bechtold T, Mussak R (eds) Handbook of natural colorants. Wiley, London

Santana ÁL, Debien ICN, Meireles MAA (2017a) High-pressure phase behavior of turmeric waste and extracts in the presence of carbon dioxide, ethanol and dimethylsulfoxide. J Supercrit Fluids 124:38–49

Santana ÁL, Meireles MAA (2016) Thin-layer chromatography profiles of non-commercial turmeric (Curcuma longa L.) products obtained via partial hydrothermal hydrolysis. Food Publ Health 6:15–25

Santana ÁL, Meireles MAA (2017) Coprecipitation of turmeric extracts and polyethylene glycol with compressed carbon dioxide. J Supercrit Fluids 125:31–41

Santana ÁL, Osorio-Tobón JF, Cardenas-Toro FP, Steel CJ, Meireles MAA (2017b) Partial-hydrothermal hydrolysis is an effective way to recover bioactives from turmeric wastes. Food Sci Technol (Campinas) Submitted

Santana ÁL, Zabot GL, Osorio-Tobón JF, Johner JCF, Coelho AS, Schmiele M, Steel CJ, Meireles MAA (2017c) Starch recovery from turmeric wastes using supercritical technology. J Food Eng 214:266–276

Santos DT, Meireles MAA (2013) Micronization and encapsulation of functional pigments using supercritical carbon dioxide. J Food Process Eng 36:36–49

Sherma J, Fried B (2003) Handbook of thin-layer chromatography, 3 edn

Shrivastava A, Gupta VB (2011) Methods for the determination of limit of detection and limit of quantification of the analytical methods. Chronicles Young Sci 2:21–25

Soponar F, Moţ AC, Sârbu C (2008) Quantitative determination of some food dyes using digital processing of images obtained by thin-layer chromatography. J Chromatogr A 1188:295–300

Świeboda R, Jóźwiak A, Jóźwiak G, Waksmundzka-Hajnos M (2014) Thin-layer chromatography and chemometric studies of selected potentilla species. Amer J Anal Chem 5:1109–1120

Xu C, Lancaster J (2008) Conversion of secondary pulp/paper sludge powder to liquid oil products for energy recovery by direct liquefaction in hot-compressed water. Water Res 42:1571–1582

Xu Y, Cai F, Yu Z, Zhang L, Li X, Yang Y, Liu G (2016) Optimisation of pressurised water extraction of polysaccharides from blackcurrant and its antioxidant activity. Food Chem 194:650–658

Zhang L (2006) Background reconstruction for camera-based thin-layer chromatography using the concave distribution feature of illumination. Opt Express 30:10386–10392

Chapter 4
Perspectives on Vanillin Production from Sugarcane Bagasse Lignin Using Supercritical CO_2 as a Solvent in a Novel Integrated Second-Generation Ethanol Biorefinery

Abstract The use of supercritical CO_2 (SC–CO_2) as a solvent for the extraction of vanillin from organosolv media was proposed and evaluated. The use of sugarcane bagasse as a biomass source to product diversification has been gaining much attention recently and it is a very promising research topic. From sugarcane bagasse, it is already produced fuels as 2G ethanol and others are being investigated as methanol, SNG. Also, different bioproducts as PET area already in the market and bioproducts as xylitol, vanillin, etc. are under study or investigated in pilot plants. Thus, in this chapter, some perspectives on vanillin production from sugarcane bagasse lignin using supercritical CO_2 as a solvent in a novel integrated second-generation ethanol biorefinery are presented.

4.1 Introduction

Biomass is a relatively abundant, renewable, carbon-neutral material that can be quite easily used to substitute fossil fuels. The very promising concept of using biomass as a raw material to produce a big variety of products ranging from fuels to chemicals and polymers is called a biorefinery. From the three polymers that constitute lignocellulosic biomass, cellulose, hemicelluloses, lignin, usually the cellulose and hemicellulose streams have well-developed processes to be valorized (Albarelli et al. 2013).

Biomass fractionation techniques to envision a full use of biomass to bioproduction envisioning a real biorefinery concept were fuels, chemicals, and high added-value compounds, as food additives and pharmaceutical/nutraceuticals, could be produced. The sugarcane sector moves a big share of the Brazilian economy. This industry conventionally produces sugar, ethanol, and electricity. Nevertheless, in the last decades, it has undergone a major modernization to optimize the production process and envisions increasing biofuel and other bioproducts production using process by-products and/or process wastes (Santos et al. 2014).

The use of bagasse as a biomass source for product diversification is very promising. From sugarcane bagasse, it has already produced fuels as 2G ethanol and others are being investigated as methanol, SNG. Also, different bioproducts as PET area

D. T. Santos et al., *Supercritical Fluid Biorefining*,
SpringerBriefs in Applied Sciences and Technology,
https://doi.org/10.1007/978-3-030-47055-5_4

are already in the market and bioproducts as xylitol, vanillin, etc. are under study or investigated in pilot plants (Mian et al. 2014).

In general, the lignin stream obtained is rarely valorized and if it is at all used, it is only used as a low-grade boiler fuel for heating or power generation purposes. Lignin can be used as a precursor of both vanillin and syringaldehyde, which are currently mainly produced by petrochemical routes. Vanillin is a flavoring and fragrance ingredient, broadly used in food and fragrances industries but can also be used for drug and chemicals synthesis. Up to now, there exist two types of vanillin: the natural and the synthetic ones, but because of its very high price, the natural's vanillin market is very small when compared to synthetic one (4 t/year compared to 16,000 t/year). The biggest share of synthetic vanillin is produced through the catechol-guaiacol route (85%), which is highly dependent on the petroleum price, while the rest is already produced by utilizing one type of lignin (lignosulfonates). Vanillin can be produced from the alkaline oxidation of lignin by using as an oxidizing agent: molecular oxygen, air, nitrobenzene or metallic oxides.

In this context, this chapter evaluates the study of valorizing lignin to products. Many studies have been evaluated different biorefinery concepts to production of biofuels from sugarcane (Albarelli et al. 2014). But, until now only the Organosolv process was evaluated, which only resulted in low-grade lignin, then and it was only considered as fuel to the cogeneration system. With the evaluation of the Organosolv process described by Wallerand et al. (2014), high-quality lignin was produced and with the aid of supercritical fluid extraction (SFE) of vanillin using CO_2 vanillin was recovered. Vanillin was chosen as a potential bioproduct and a mathematical simulation of the product was accomplished. The model was improved, and the idea is to evaluate this process integrated into a theoretical sugarcane biorefinery. The description of the biorefinery and the vanillin production process is given in the following.

4.2   Materials and Methods

4.2.1   Biorefinery of Sugarcane Considered to the Production of High-Quality Lignin

The biorefinery considers the production of ethanol from sugarcane juice and sugarcane bagasse cellulose, the production of xylose to xylitol production and the production of high-quality lignin that could be used to vanillin production. Figure 4.1 shows the overall sugarcane biorefinery considered. For more details on the parameters used for simulation see (Wallerand et al. 2014).

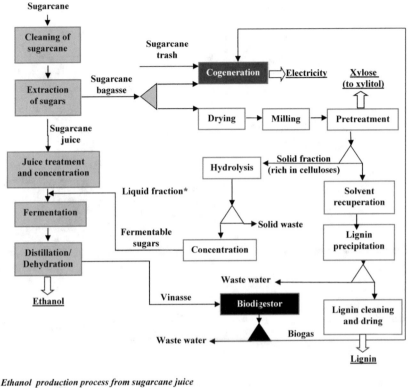

Fig. 4.1 Sugarcane biorefinery

4.2.2 Vanillin Production Process

The production of vanillin consists of two main blocks of processes: the reaction block and the separation block. Figure 4.2 shows the hierarchy blocks used in Aspen to simulate the process. The detailed information on each block is given in the following.

Reaction Block

In order to design the simulation process for the reaction stage, the studies of Sales et al. (2003, 2007). The process followed in order to transform lignin to vanillin, syringaldehyde, and p-hydroxybenzaldehyde is oxidation (using high-pressure molecular oxygen) performed in an alkaline environment (NaOH(2 M)) under the presence of a catalyst ($PdCl_3.3H_2O/g–Al_2O_3$). The reactor conditions applied are described in Table 4.1.

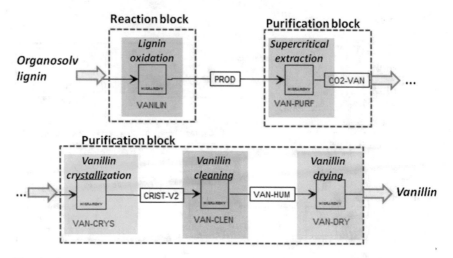

Fig. 4.2 Hierarchy blocks used in aspen to simulate the vanillin production process from organosolv lignin

Table 4.1 Parameters adopted in the reaction step

Reaction parameters	Data
Catalyst	4 wt%
Temperature	140 °C
Total pressure	20 bar
Oxygen pressure	5 bar
Concentration of lignin	60 g/L
Concentration of NaOH	2 M
Conversion yields	
Vanillin	4.51%
Syringaldehyde	4.64%
p-Hydroxybenzaldehyde	7.09%

Based on Sales et al. (2003, 2007)

The reaction system can be described by the following set of reactions:
Vanillin production

$$C_{10}H_{13.9}O_{1.3} + 4.325O_2 \rightarrow C_8H_8O_3 + 2CO_2 + 2.95H_2O$$

Syringaldehyde production

$$C_{10}H_{13.9}O_{1.3} + 3.325O_2 \rightarrow C_9H_{10}O_4 + CO_2 + 1.95H_2O$$

Table 4.2 Main parameters assumed in the supercritical fluid extraction step

Reaction parameters	Data
Temperature	60 °C
Total pressure	125 bar
CO_2 flow	600 g/g of vanillin
Vanillin loss	2%
Vanillin concentration in the extract[a]	90%

[a]The remaining percentage corresponds to the other compounds, Syringaldehyde, etc.

Based on Klemola and Tuovinen (1989)

p-Hydroxybenzaldehyde production

$$C_{10}H_{13.9}O_{1.3} + 5.325O_2 \rightarrow C_7H_6O_2 + 3CO_2 + 3.95H_2O$$

Remaining lignin reaction

$$C_{10}H_{13.9}O_{1.3} + 12.825O_2 \rightarrow 10CO_2 + 6.95H_2O$$

Separation Block

The separation block consists of the post-processing of the products to attain the final vanillin product. In the reaction block, there are four steps, the supercritical fluid extraction (SFE) of vanillin using CO_2, the crystallization, the cleaning, and the vanillin drying.

Supercritical fluid extraction (SFE): This method consists of the extraction of vanillin by using CO_2 under supercritical conditions ($P = 75$–400 bar). During that process, vanillin is dissolved in the CO_2 stream and afterwards is recovered by passing the flow through a receiver. Advantages of this system are the very high level of purities (90% compared to 60% of conventional extraction stages) obtained but also the handling CO_2, a nontoxic and easy to handle substance instead of increased amounts of acids and solvents which are otherwise required (Table 4.2).

Crystallization: This process was considered to increase the vanillin purity to 97.54% (Klemola and Tuovinen 1989) and prepare the product to be sold at its solid form. Table 4.3 shows the main parameters considered to simulate this step.

Vanillin cleaning and drying: The final step considered was to clean and dry the vanillin crystals. Table 4.3 shows the main parameters considered to simulate this step.

4.3 Results and Discussion

The solution which was obtained from the vanillin production stage still contains unreacted lignin, water, Na^+, both in the form of the aldehydes' salts and in NaOH excess solution, and a mixture of aldehydes. Water is not soluble in CO_2, thus CO_2 progressively decrease the pH of the solution which will lead in the dissociation of the

Table 4.3 Main parameters considered to simulate the crystallization, cleaning and drying step

Reaction parameters	Data
Crystallization step	
Crystallization medium	Water
Temperature[a]	20 °C
Pressure	1 bar
Vanillin purity[b]	97.54%
Vanillin not crystallized	30%
Water content after crystals separation	50%
Cleaning step	
Water flow	2 times the vanillin flow
Water content after centrifugation	50%
Drying step	
Final moisture	2.2%

[a]Chosen base on (Tarabanko et al. 2013)
[b]Based on Klemola and Tuovinen (1989)
The remaining parameters were assumed as no literature data was found

aldehydes salts and the formation of the respective aldehydes (vanillin, syringaldehyde, and p-hydroxybenzaldehyde) and also the unreacted fragments of lignin are not extracted and therefore remain in the oxidation solution. Additionally, similar products to vanillin, such as the remaining aldehydes and other phenolic compounds can be extracted with the supercritical CO_2 stream and recovered afterward as a mixture that can be easier separated.

The SFE process requires 200–800 kg of supercritical CO_2 for one kilo of vanillin. Subsequently, since all aldehydes of the mixture react in the same way, 200–800 kg will be needed for one weight of the sum of the three aldehydes (530.8 kg). It is therefore expected that the main operation cost of the process will result from the power requirement of the relative compressor used since the above quantities will have to reach the required supercritical conditions of CO_2 (150 bar, $T > 30$ °C).

Additional treatments are needed for the recovery of the aldehydes from the CO_2-aldehydes mixture and the recovery of lignin and NaOH from the oxidation solutions. The mixture which is reached after the SFE stage contains the mixture of the three aldehydes and approximately 10% further impurities. Them it is proposed further one or two steps crystallization from water in order to obtain very high purities (96–99.5%). The mixture of the above aldehydes should be separated, but since the separation of only mixtures of vanillin and syringaldehyde (ammonia method, extraction, and crystallization) were found, a more extensive bibliographic review is proposed or the mixture can alternatively be sold as it is at a lower price, so that another actor may separate it.

4.4 Conclusions

The proposed production of vanillin from the specified lignin stream, a method which comprises of four main steps: reaction, gas/liquid separation, supercritical CO_2 extraction, and crystallization stages was proposed in order that a mixture of aldehydes (vanillin, syringaldehyde, and p-hydroxybenzaldehyde) is reached. It can be concluded that although the reaction step is quite simple, the separation steps needed in order to obtain the required purity (97% for technical applications and 99.8% for food applications) are difficult and numerous.

Supercritical fluid extraction (SFE) has demonstrated to be an ideal clean technology to be used as part of a holistic biorefinery for the selective recovery of vanillin from vanillin from organosolv media. New perspectives on how biomasses can be better valorized with the aid of SFE process have culminated in several developments aiming full use of biomass, indicating that SFE can be effectively used in novel biorefinery concepts. In particular, some advantages of SFE for this application are (i) the very high level of purities (90% compared to 60% of conventional extraction stages) obtained, but also, (ii) the handling of CO_2, which is a nontoxic and easy to handle substance instead of increased amounts of acids and solvents, which are otherwise required.

Acknowledgements Juliana Q. Albarelli thanks FAPESP (processes 2013/18114-2; 2015/06954-1) for the post-doctoral fellowships. Diego T. Santos thanks CNPq (processes 401109/2017-8; 150745/2017-6) and CAPES (process 7545-15-0) for the post-doctoral fellowships. M. Angela A. Meireles thanks CNPq for the productivity grant (302423/2015-0). The authors acknowledge the financial support from CNPq (process 486780/2012-0) and FAPESP (processes 2012/10685-8; 2015/13299-0).

References

Albarelli JQ, Ensinas AV, Silva MA (2013) Product diversification to enhance economic viability of second generation ethanol production in Brazil: the case of the sugar and ethanol joint production. Chem Eng Res Des 92:1470–1481

Albarelli JQ, Onorati S, Caliandro P, Peduzzi E, Ensinas AV, Maréchal F (2014) Thermo-economic optimization of integrated ethanol and methanol production in the sugarcane industry. Chem Eng Trans 39:1741–1746

Klemola A, Tuovinen J (1989) Method for the production of vanillin, United States patent 4847422

Mian A, Albarelli JQ, Ensinas AV, Maréchal F (2014) Integration of supercritical water gasification in combined 1G/2G ethanol production. Chem Eng Trans 39:1795–1800

Sales FG, Abreu CAM, Pereira JAFR (2003) Catalytic wet-air oxidation of lignin in a three-phase reactor with aromatic aldehyde production. Braz J Chem Eng 21:211–218

Sales FG, Maranhao LCA, Filho NML, Abreu CAM (2007) Experimental evaluation and continuous catalytic process for fine aldehyde production from lignin. Chem Eng Sci 62:5386–5391

Santos DT, Albarelli JQ, Rostagno MA, Maréchal F, Meireles MAA (2014) New proposal for production of bioactive compounds by supercritical technology integrated to a sugarcane biorefinery. Clean Technol Environ Policy 16:1455–1468

Tarabanko VE, Chelbina YV, Kudryashev AV, Tarabanko NV (2013) Separation of vanillin and syringaldehyde produced from lignins. Sep Sci Technol 48:127–132

Wallerand AS, Albarelli JQ, Ensinas AV, Ambrosetti G, Mian A, Marechal F (2014) Multi-objective optimization of a solar assisted 1st and 2nd generation sugarcane ethanol production plant, Proceedings of ECOS

Chapter 5
Novel Biorefinery Concept for the Production of Carotenoids from Microalgae Using Lignocellulose-Based Biorefinery Products and Supercritical Fluids

Abstract This chapter proposes an innovative biorefinery conceptual process for the production of carotenoids from microalgae using lignocellulose-based biorefinery products and/or by-products and pressurized fluids. The extraction process, which can be done also with microalgal biomass with high content of moisture avoiding high-cost downstream processes, involves the use of ethanol and 2-MethylTetraHydroFuran (2 MTHF) mixed or in a sequential form for selective extraction of carotenoids. 2 MTHF is obtained from furfural, which is produced as a by-product during lignocellulosic biomass (sugarcane bagasse, wood, corn stover, rice straw, etc.) pretreatment for ethanol production, for example. The solvent recovery step involves the use of CO_2, which is obtained from ethanol fermentation as a by-product. Specific conditions for CO_2 for temperature and pressure to achieve supercritical conditions would be applied in order to besides high solvent recovery and recycling provide a desirable selective carotenoid purification and encapsulation if a coating material is added. The two-step process can be converted in a one-step process minimizing carotenoid degradation if the extraction process is performed under higher pressure than that performed during extract precipitation. In addition, the proposed processing route can be well integrated into conventional existing biofuels production (gasification, combustion, etc.) scenarios using the solids recovered after carotenoids production as feedstock.

5.1 Introduction

The complexity of natural product matrices, the need for isolating specific bioactive compounds, and the high costs involved are promoting the development of new strategies to improve the whole process. One of these developments is the concept of integrating different stages into one single on-line operation (Rostagno et al. 2010).

Usually, several different processes are required for the production of highly concentrated extracts. The most important processes involved in the production of extracts from natural products are the extraction of target compounds, their purification, and the elimination of the solvent and their stabilization by encapsulation and particle formation. Most of these processes are performed sequentially and one

© The Author(s), under exclusive license to Springer Nature Switzerland AG 2020
D. T. Santos et al., *Supercritical Fluid Biorefining*,
SpringerBriefs in Applied Sciences and Technology,
https://doi.org/10.1007/978-3-030-47055-5_5

process cannot start before the preceding has been completed (off-line processing). Depending on the raw material and on the desired characteristics of the final product, different processes, and techniques may be used.

As it is a complex process with several factors involved, the production of extracts from natural sources can be a challenge. The high operational and investment costs involved, due to several pieces of equipment, long process time, and associated labor and utilities, are translated to the manufacturing costs of the extracts, which is one of the main challenges to be faced by the industry (Prado et al. 2011). In this context, the use of process integration can be explored to address these problems and set the basis for a modern and efficient natural product industry. However, it is necessary to have adequate knowledge of the processes involved in order to explore their characteristics at most and to improve the overall process.

The processing of most natural products for the production of extracts involves the use of one or more solvents (liquid, supercritical or a mixture of both) in a sequential manner using different processes. Therefore, it presents a high potential to implement the concept of process integration. In most cases, pressurized fluid technology (pressurized liquids and sub/supercritical fluids) can be used to replace traditional methods and fully integrate the processes from extraction to solvent evaporation and particle formation. In fact, pressurized fluid technology is not considered as an alternative to single procedures; the full potential of this technology can only be achieved by using an integrated approach (King and Srinivas 2009; Temelli 2009). On the other hand, there is only a few reports are available on that subject.

5.2 Prior Art Searches

5.2.1 Integral Use of Algal Biomass

The past decade has seen a surge in the interest in microalgae culture for renewable fuel production as an alternative to petroleum transport fuels. Microalgae are currently considered a promising feedstock for next-generation renewable fuels production because of their much higher photosynthetic efficiency, areal productivity, and lipid content and also that they do not compete with food cultures. Another advantage is that the growth of microalgae can transport atmospheric carbon into a cycle in which no additional CO_2 is created (Gouveia and Oliveira 2009; Yen et al. 2013; Brennan and Owende 2010).

Microalgae contain valuable compounds such as lipids, pigments, proteins, and carbohydrates, which all can be used for different markets. Nevertheless, there are still several problems to be solved during the development of economically feasible microalgae-based biorefinery concepts. Some biorefinery concepts for microalgae use is under development (Pignolet et al. 2013; Vanthoor-Koopmans et al. 2013). Lipid and carotenoid compounds are being considered as a unique product or sometimes even the pigment content in the oil product has being negligenced. Meanwhile,

Supercritical Fluid Extraction (SFE) method has being pointed out as the most technical promising option for this fractionation step due to the reduced disposal of organic solvent (Palavra et al. 2011; Ruen-ngam et al. 2012).

Since to perform SFE process it is necessary to remove the water after harvesting biomass, which consumes tremendous amount of energy, currently most of the efforts are devoted to developing processes for producing products directly from wet algal biomass. Lipid-rich extracts methods have been successfully developed by the direct extraction from wet microalgae (Du et al. 2013; Yoo et al. 2012; Halim et al. 2011). Among them, the Pressurized Liquid Extraction (PLE) method has demonstrated to be a promising alternative in order to save energy consumption and increase the extraction yield of lipids (Chen et al. 2012), but none of these works have studied the fractionation of the extracts in order to separate lipids from carotenoids neither to selectively recovery specific carotenoid type. In addition, none of these works have studied the selective extraction of the carotenoids leaving after the extraction step a solid residue rich in valuable compounds that can be a suitable feedstock for biorefining.

The main bottleneck is to separate the different fractions selectively without damaging one or more of the product fractions. Sustainable technologies to overcome these bottlenecks need to be developed. In this context, due to PLE method allow for fast extraction and reduced solvent consumption (Chen et al. 2012; Santos et al. 2012a, b) we are developing a process using this technique using renewable solvents, ethanol, and 2-MethylTetraHydroFuran (2-MTHF) mixed or in a sequential form, for selective extraction of carotenoids directly from wet microalgae, avoiding high-cost downstream processes. 2-MTHF is derived from renewable resources such as waste lignocellulosic biomass and to the best of our knowledge, this is the first research study that uses this solvent, which has increasingly being used in chemical and biochemical reactions as reaction medium (Aycock 2007; Gao et al. 2013), was used as extraction solvent of carotenoids.

Major components of microalgae biomass are lipids (10–20% w/w), proteins (50–60% w/w) and carbohydrates (10–15%). Normally, the lipids are pointed out the most important fraction when considering the potential for biofuels production since the production of biodiesel by transesterification is a well-known and mature technology, being all the other fractions been lost or degraded during the conversion process. Besides the loss of economic value when microalgae are directly used as feedstock for biofuels conversion other disadvantages are present since burning microalgae containing a high amount of protein and consequently, nitrogen leads to the formation of greenhouse gases, like NO_x, during the conversion process.

Therefore, it is expected that the innovative process for the production of carotenoids from microalgae that is being developed besides carotenoids extraction using environmentally friendly solvents has the benefit of producing a solid product containing high amounts of proteins, lipids, and carbohydrates that can be further recovered for different uses, enhancing its economic value.

5.2.2 Sequential Extraction Using Pressurized Fluids

Sequential extraction is a well-known procedure that can be useful to improve the process selectivity and the recovery of different types of extracts from the same raw material. Depending on the raw material, it is possible to perform successive extractions employing different solvents or process conditions (pressure and/or temperature) to selectively extract different classes of compounds. Pressurized liquids and supercritical fluids present several characteristics that can be explored to achieve this objective.

In general, supercritical CO_2 can be employed for extracting nonpolar to moderately polar phytochemicals, while water, ethanol, and other polar organic solvents are better for extracting polar compounds. For the extraction of moderately polar phytochemicals using supercritical CO_2, it is usually necessary to add a cosolvent, such as ethanol or another organic solvent. The amount and type of cosolvent depend on the matrix and the polarity of the target compounds, with concentrations ranging from 1 to 90% of the total solvent mass (Pereira and Meireles 2010).

There are several reports available in the literature where a sequential extraction strategy was adopted for a comprehensive extraction of different compound classes from the same raw material. In several cases, supercritical fluid extraction (SFE) using CO_2 is employed in a first extraction step to extract lipophilic compounds and later a more polar solvent (liquid solvent or modified CO_2) also under pressure is employed in order to extract polar compounds (Table 5.1).

To illustrate the principle we can take as an example a study found in the literature dealing with grape seeds. In this case, the grape seeds were initially extracted using pure supercritical CO_2, which removed over 95% of the oil present. In a sequential extraction step, the residue was re-extracted using subcritical CO_2 modified with methanol (40%); this step removed over 79% of catechins and epicatechins present. In the last extraction step, polyphenolicdimers/trimers and procyanidins were extracted from the residue using pure methanol. Each extraction step was carried out using different conditions and produced a different extract with unique composition and characteristics. Furthermore, the whole process was carried out on a single instrumental extraction system, demonstrating the potential of pressurized fluid technology to implement the sequential extraction strategy (Ashraf-Khorassani and Taylor 2004).

This is only one of the possible approaches to adopt this strategy, but depending on the raw material different processes can be combined or replaced. The main interest in the use of the sequential extractions is to fully explore the potential of the raw material to produce different types of extracts. But it can also be used as a tool to eliminate undesirable components of the raw material and to improve extraction yields of target compounds. The extraction of grape skins with supercritical CO_2, for instance, improved the subsequent recovery of polyphenols from the residue using 50% ethanol–water mixture at 60 °C under atmospheric pressure (Vatai et al. 2009). The removal of nonpolar components by the CO_2 increased the yields of the second step by 2–3 times when compared to the single-step conventional extraction.

Table 5.1 Applications of sequential processes for the extraction and purification of natural products

Raw material	Components/compounds	Process	Observations	References
Jabuticaba (*Myrciaria cauliflora*)	Anthocyanin pigments and lipophilic compounds	*First step* PLE using ethanol to obtain polar compounds including anthocyanin pigments *Second step* SFE with CO_2 to recover low polarity compounds	Fractionated extractions of jabuticaba skins were successfully performed, producing two valuable extracts with antioxidant activities. The extract from the first step was rich in anthocyanin pigments, and the extract from the second step was rich in lipophilic compounds including essential oils and less polar flavonoid compounds	Santos et al. (Santos et al. 2011)
Turmeric (*Curcumalonga* L.)	Curcuminoids	*First step* SFE using CO_2 at 22.5 MPa and 35 °C *Second step* SFE using CO_2 and 50% ethanol or isopropyl alcohol at 30 MPa and 30 °C	The SFE allowed to obtain a volatile oil fraction using pure CO_2 and a curcuminoids fraction using cosolvents	Braga and Meireles (2007)

(continued)

Table 5.1 (continued)

Raw material	Components/compounds	Process	Observations	References
Jambu flowers, leaves and stems (*Spilanthesacmella*)	Spilanthol	*First step* SFE using CO_2 as solvent at 25 MPa and 50 °C *Second step:* ESE using CO_2 as solvent and ethanol, water and their mixtures as cosolvent at 25 MPa and 50 °C *Third step* ESE using methanol	The first step extracted most of the spilanthol while the second step removed only small amounts of spilanthol that still remained in the vegetal matrix. Higher extraction yields, total phenolic compounds, and compounds with high antioxidant activity were obtained when using organic/polar solvents as enhancers, as was the case of ESE (H_2O) and ESE (EtOH + H_2O) from flowers and ESE (H_2O) from leaves	Dias et al. (2012)

(continued)

Table 5.1 (continued)

Raw material	Components/compounds	Process	Observations	References
Chardonnay grape seeds	Oils, polyphenols, and procyanidins	*First step* SFE using CO_2. *Second step* SFE of the extract obtained in the first step using CO_2 with modifier (ethanol) under different conditions of temperature (50–80 °C), pressure (10–30 MPa) and concentration of modifier (1–5%)	Pure supercritical CO_2 removed over 95% of the oil from the grape seeds. Subcritical CO_2 modified with methanol extracted monomeric polyphenols, whereas pure methanol extracted polyphenolic dimers/trimers and procyanidins. At optimum conditions, 40% methanol modified CO_2 removed 79% of catechin and epicatechin from the grape seed. The third step provided a dark red solution shown via electrospray ionization HPLC-MS to contain a relatively high concentration of procyanidins	Ashraf-Khorassani and Taylor (2004)

CO_2 Carbon dioxide; ESE enhanced solvent extraction; EtOH ethanol; H_2O water; PLE pressurized liquid extraction; SFE supercritical fluid extraction

Obviously, the sequential extraction strategy can be used for off-line processing or for combining techniques. However, combined processes using pressurized fluids in both extraction steps can take advantage of automation and on-line control of the process (Zougagh et al. 2004). Additionally, there is no material discharge after the first extraction, as it allows both extractions to be carried out in the same vessel eliminating a unit operation, thereby reducing costs.

There are many possibilities for the application of the process integration concept. It can be assumed that the adequacy and economic feasibility of the implementation of a sequential extraction scheme is basically determined by the raw material used, the products obtained and additional costs associated with the use of two or more extraction steps. Therefore, from an economic perspective, it is necessary to balance the manufacturing costs with the expected benefits for each product.

5.2.3 *Integration of the Extraction Process to Purification and/or Encapsulation Processes*

In general terms, the main goal of an extraction and purification process is to produce a highly concentrated extract rich in specific compounds or compound classes, which can be used directly as an additive or by itself as a product. However, some compounds present in the extracts are unstable under certain conditions and may be subject to transformations or degradation after the extraction, hindering their utilization industrially. Therefore, stabilization of extracts where such compounds are present is especially important to ensure the desired activity/property when the additive/product is actually used or consumed.

In this aspect, protective techniques can be used for the stabilization of natural extracts and for the protection of sensitive compounds from moisture, oxidation, heat, light or extreme conditions during processing, in an effort to increase their shelf life and range of applications. Furthermore, these techniques can be used to mask undesirable component attributes, such as strong and unpleasant flavors, attending to sensory quality and functionality and to promote the controlled release of the active component (Fernandez and Torres-Giner 2009; Santos and Meireles 2010).

Stabilization techniques are becoming an essential tool to increase the competitiveness of natural products and to allow their effective use by the industry, helping to increase shelf life and protecting the properties/activity of the encapsulated material (Semeonova et al. 1992).

There are several encapsulation techniques available, which can be classified according to the process of combination between coating and core material into three categories: physical, chemical and physicochemical processes. Furthermore, the particles formed by these encapsulation techniques may be classified according to their size in macro (>5000 μm), micro (1.0–5000 μm), and nanoparticles (<1.0 μm) (Jafari et al. 2008).

Although several of these techniques are currently being used industrially, it is noteworthy that all of them have inherent limitations. These include poor control of particle size and morphology, degradation of thermosensitive compounds and low encapsulation efficiency. These limitations are prompting the development of new techniques and several different processes are currently being used for natural products in replacement of conventional encapsulation processes (Santos and Meireles 2010).

Supercritical fluid technology is an alternative to conventional encapsulation techniques which allows obtaining solvent-free micro/nanoparticles and capsules with narrow size distribution. Carbon dioxide is the primary fluid applied to produce composite particles using supercritical fluid methods because it enables the process to be performed at near ambient temperature in an inert atmosphere, which avoids the degradation of the sensitive compounds by heat and oxygen. The supercritical state of carbon dioxide is achieved at moderate pressures and temperatures (31 °C and 7.38 MPa, respectively), which is suitable for most applications. Because of these advantages, there are several encapsulation techniques employing supercritical technology. The supercritical fluids encapsulation techniques can be distinguished from each other according to the role of the supercritical fluid in the process (Domingo et al. 2003):

1. as a solvent: rapid expansion of supercritical solutions (RESS); supercritical solvent impregnation (SSI);
2. as a solute: particles from gas saturated solutions (PGSS);
3. as an antisolvent: supercritical antisolvent (SAS); supercritical fluid extraction of emulsions (SFEE).

One of the most researched processes is SAS, in which the solute of interest is first dissolved in a conventional organic solvent and then the solution is sprayed continuously into a chamber through a coaxial nozzle co-currently with the supercritical CO_2. The high-pressure CO_2 acts as an antisolvent, decreasing the solubility of the solutes in the solvent mixture. Therefore, a fast supersaturation takes place, leading to nucleation and formation of micro or nanoparticles (Sosa et al. 2011). If a coating material is also dissolved in the organic solvent, encapsulates are formed by coprecipitation with the solute (Bahrami and Ranjbarian 2007; Cocero et al. 2009). The SAS process has several figures of merit, including lower operating temperature than that used in conventional processes such as spray drying, and lower residual solvent in the final product. Also, mean particle size, particle size distribution, and morphology can be controlled by changing process parameters such as pressure and temperature (Guha et al. 2011). This process is increasingly being used to produce micro to nanometer-sized and encapsulated extracts from natural products. Examples of successful applications of this process for the encapsulation of natural products include extracts produced from green tea, rosemary (*Rosmarinus officinalis*), Annatto (*Bixa orelana* L.) and pink shrimp (*P. brasiliensis* and *P. paulensis*) residue among others (Table 5.2).

In the case of off-line processing, the efficiency of the SAS process is independent of the technique used for the production of the extract. As can be seen in Table 5.2,

Table 5.2 Applications of SAS process to produce encapsulated extracts

Raw material	Application	Process	References
Green tea (*Camellia sinensis*)	Encapsulation of green tea extract	Extraction method: microwave-assisted extraction (MAE) with acetone SAS: coating material: polycaprolactone Pressure: 8–12 MPa Temperature: 10–34 °C	Sosa et al. (2011)
Rosemary (*Rosmarinus officinalis*)	Encapsulation of rosemary extracts	Extraction method: Soxhlet with methanol SAS: carrier material: polycaprolactone Pressure: 20–30 MPa Temperature: 40 °C Flow rate: 20 g/min	Yesil-Celiktas and Cetin-Uyanikgil (2012)
Annatto (*Bixa orelana* L.)	Encapsulation of bixin-rich extract from annatto seeds	Extraction method: supercritical fluid extraction with CO_2 at 31 MPa and 60 °C SAS: coating material: Polyethylene glycol (PEG) Pressure: 10 MPa	Santos and Meireles (2013)
Pink shrimp (*P. brasiliensis* and, *P. paulensis*) residue (waste from shrimp processing)	Pink shrimp (*P. brasiliensis* and, *P. paulensis*) residue (waste from shrimp processing)	Extraction method: solvent extraction (maceration) with acetone SAS: coating material: Pluronic F127 Pressure: 8–12 MPa Temperature: 35–45 °C	Mezzomo et al. (2012)

the process can be applied to the encapsulation of extracts produced by any extraction technique and using different encapsulation agents and organic solvents. However, some aspects should be considered in order to explore the full potential of this technique. Besides the compatibility between the organic solvent, the extract and the coating agent, it is important to have adequate solvent evaporation by the supercritical CO_2. Although the extraction solvent can be eliminated and the extract can be redissolved in another more suitable solvent, this is not practical or even logical and may increase manufacturing costs due to high energy consumption. Therefore, the logical approach is to use the same solvent for the extraction and for the encapsulation of the extract. This is an extremely important aspect for on-line processing, where the type and amount of extraction solvent will influence the encapsulation and particle formation process.

The development of coupled processes for combining bioactive compound extraction to on-line particle formation is a recent trend and only a few reports are available on that subject. There is a recent report on the development of an on-line process to obtain dried powders of extracts from natural sources in one single operation (Ibanez et al. 2009). This patented process was defined by the authors as water extraction and particle formation on-line (WEPO). As the name implies, this process employs water as extraction solvent. In this case, supercritical CO_2 is not suitable for the elimination of the extraction solvent due to the low solubility of water in CO_2. In general, supercritical CO_2 is used as a dispersion medium and a hot N2 stream is used as the drying agent. A similar on-line process was more recently developed, where PLE and particle formation are coupled, was also reported using organic solvents as extraction solvent instead of water. Due to the similarities with the WEPO process, this process was defined as organic solvent extraction and particle formation on-line (OEPO) (Santos et al. 2012a, b). Differently from the WEPO process, the OEPO process allows the encapsulation of the extract immediately after its production. Indeed, the OEPO process consists of coupled PLE-SAS precipitation, PLE-SAS co-precipitation, and PLE-SFEE. The results demonstrated that this novel process, not patented, can be considered as a suitable and promising process to obtain, in only one step, different products (precipitated extract, co-precipitated extractor encapsulated extract in suspension) with desired particle size directly from the plant material. Until now, only dichloromethane and ethyl acetate were tested as extraction solvents and few bioactive compounds sources (annatto seeds and Brazilian ginseng roots) were treated with this on-line process.

5.2.4 Integration of New Processes into Existing Industrial Facilities

The integration of new processes into an existing industrial facility that generates the required resources is a likely approach that can minimize the environmental impact while simultaneously benefiting the already existing industry. The production process of carotenoids and other bioactive compounds from vegetable sources is normally a stand-alone process. On the other hand, depending on the required resources for this process, the co-location of it with an existing industrial facility seems very promising.

Sugarcane biorefinery has been already identified as a potential source for CO_2, vinasse, and energy for microalgae production (Lohrey and Kochergin 2012). It was also identified by some researchers the advantages of constructing an extraction unit that uses CO_2 as solvent and ethanol as cosolvent for the recovery of bioactive compounds from vegetable sources in close proximity to it (King and Srinivas 2009). On the other hand, to the best of our knowledge, the proposed approach that includes the integration of a one-step process for carotenoid recovery from microalgae, encapsulated or not, using lignocellulose-based biorefinery products was not previously identified. In the proposed integrated processing route ethanol and/or 2 MTHF would be used mixed or in a sequential form as the extraction solvent and CO_2 as an antisolvent during purification and/or encapsulation of the extracts obtained. In addition, the proposed processing route does not require algal biomass drying, which has been pointed out as the main bottleneck for alga biomass processing and allows the integral use of this biomass.

5.3 Description of the Proposed Supercritical Fluid Biorefinery Concept

Figure 5.1 shows the proposed processing route integrated into a lignocellulose-based biorefinery using sugarcane bagasse as feedstock producing several products. And Fig. 5.2 shows the scheme of the proposed one-step process for carotenoid-rich extract production encapsulated or not.

The one-step process for carotenoid-rich extract production encapsulated or not combines the two different processes previously described: first a dynamic pressurized liquid extraction (PLE) process using ethanol and/or 2-MethylTetraHydroFuran (2 MTHF) and secondly the elimination of the extraction solvent by the precipitation of the extract, using supercritical CO_2 as an antisolvent. Thus, extraction and precipitation take place in the same system with a small time delay between these two processes.

From the extraction cell, the extract solution can be added directly to the precipitation vessel or be led to a T-mixer where it can be mixed with a solution containing a coating material dissolved in an adequate organic solvent. Shortly afterward the solution is exited through the coaxial annular passage of the atomizer together with supercritical CO_2 into the precipitation vessel.

The extraction cell can be filled with microalgal biomass with any water content. The amount inserted of plant material is calculated in order to keep the required solvent volume to feed volume ratio for adequate extraction of the bioactive compounds during the PLE experiments. The process started with a static extraction period (selected after optimization) by filling the cell with the desired extraction solvent at the desired temperature (selected after optimization) and pressure (~10 MPa). At the same time, CO_2 is pumped through the system at the desired temperature (~40 °C) and pressure (10 MPa), with a constant flow rate. The extraction continued in a continuous flow mode (dynamic extraction period) by opening valve 5 and setting the extraction solvent rate at the desired constant value. The extract solution can meet first the solution containing a coating material dissolved in an organic solvent pumped in parallel if an encapsulation product is aimed. Afterward the organic solvents from the extract solution are exited through the vessel precipitating the product

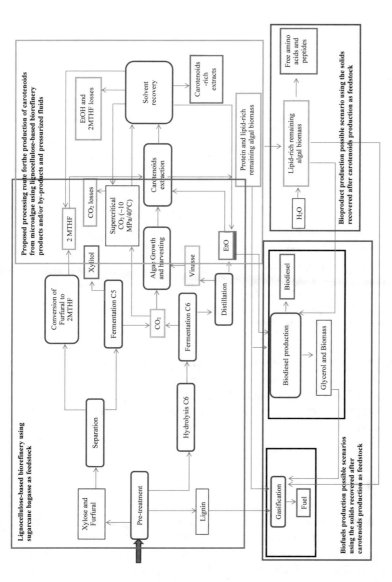

Fig. 5.1 The proposed processing route integrated into a lignocellulose-based biorefinery using sugarcane bagasse as feedstock producing as intermediate products (blue line boxes): 2-MethylTetraHydroFuran (2 MTHF) from furfural, part of ethanol (EtOH) from the fermentation of the hexoses (C6), protein and lipid-rich remaining algal biomass and lipid-rich remaining algal biomass; as final products (red line boxes): xylitol from the fermentation of the pentoses (C5), part of ethanol (EtOH) from the fermentation of the hexoses (C6), carotenoid-rich extracts, free amino acids, and peptides and biofuels; and as by-products (green line boxes): CO_2 and vinasse

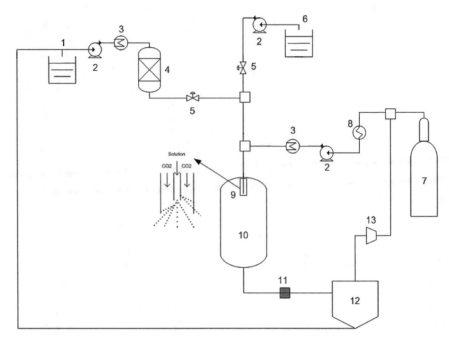

Fig. 5.2 Scheme of the proposed one-step process for carotenoid-rich extract production encapsulated or not. 1—Ethanol and/or 2-MethylTetraHydroFuran (2 MTHF) reservoir; 2—liquid pumps; 3—cooling system; 4—extractor vessel; 5—back pressure regulator valves; 6—coating material solution reservoir; 7—CO_2 cylinder; 8—heating system; 9—atomizer; 10—precipitation vessel; 11—filter; 12—flash tank; 13—CO_2 compressors

in a filter, which can be a: (i) precipitated extract; or (ii) encapsulated extract. Both, organic solvents and CO_2 can be recycled at the end. For the use of 4.5 g of plant material, a constant flow rate of CO_2 at 0.6 kg/h provided a dry precipitated extract with less than 50 ppm of remaining organic solvent when an extraction flow rate of 1.0 cm^3/min was employed.

5.4 Conclusions

The proposed one-step process can produce organic solvent-free products [two, three or more if different solvents or process conditions (pressure and/or temperature) are employed sequentially], with desirable carotenoid composition profiles in an encapsulated form or not. The use of other lignocellulose-based biorefinery by-products, like CO_2 and vinasse, the liquid waste effluent from the sugarcane biorefinery, for other purposes, such as for microalgae growth can be integrated into this new process. In addition, the proposed processing route can be well integrated into conventional existing biofuels production (gasification, combustion, etc.) scenarios using the solids recovered after carotenoids production as feedstock. Alternatively, employing new promising one-step production routes the remaining protein in the leftover solids after carotenoids production could be recovered prior to biofuel conversion by simultaneous extraction and hydrolysis process using subcritical water and/or this remaining solid material can be directly converted to biodiesel by simultaneous extraction and transesterification using pressurized ethanol. Depending on the water content of the material to be used for biofuel production the use of supercritical water gasification can be the desired technology.

Acknowledgements Diego T. Santos thanks CNPq (processes 401109/2017-8; 150745/2017-6) and CAPES (process 7545-15-0) for the post-doctoral fellowships. M. Angela A. Meireles thanks CNPq for the productivity grant (302423/2015-0). The authors acknowledge the financial support from CNPq (process 486780/2012-0) and FAPESP (processes 2012/10685-8; 2015/13299-0).

References

Ashraf-Khorassani M, Taylor LT (2004) Sequential fractionation of grape seeds into oils, polyphenols, and procyanidins via a single system employing CO_2-based fluids. J Agric Food Chem 52:2440–2444

Aycock DF (2007) Solvent applications of 2-methyltetrahydrofuran in organometallic and biphasic reactions. Org Process Res Dev 11:156–159

Bahrami M, Ranjbarian S (2007) Production of micro-and nano-composite particles by supercritical carbon dioxide. J Supercrit Fluids 40:263–283

Braga MEM, Meireles MAA (2007) Accelerated solvent extraction and fractioned extraction to obtain the curcuma longa volatile oil and oleoresin. J Food Process Eng 30:501–521

Brennan L, Owende P (2010) Biofuels from microalgae—a review of technologies for production, processing, and extractions of biofuels and co-product. Renew Sustain Energy Rev 14:557–577

Chen M, Liu T, Chen X, Chen L, Zhang W, Wang J, Gao L, Chen Y, Peng X (2012) Subcritical co-solvents extraction of lipid from wet microalgae pastes of Nannochloropsis sp. Eur J Lipid Sci Technol 114:205–212

Cocero MJ, Martín A, Mattea F, Varona S (2009) Encapsulation and co-precipitation processes with supercritical fluids: fundamentals and applications. J Supercrit Fluids 47:546–555

Dias AMA, Santos P, Seabra IJ, Junior RNC, Braga MEM, Sousa HC (2012) Spilanthol from *Spilanthes oleracea* flowers, leaves and stems obtained by selective supercritical carbon dioxide extraction. J Supercrit Fluids 61:62–70

Domingo C, Vega A, Fanovich MA, Elvira C, Subra P (2003) Behavior of poly(methyl methacrylate)–based systems in supercritical CO_2 and CO_2 plus cosolvent: solubility measurements and process assessment. J Appl Polym Sci 90:3652–3659

Du Y, Schuur B, Samorì C, Tagliavini E, Brilman DWF (2013) Secondary amines as switchable solvents for lipid extraction from non-broken microalgae. Bioresour Technol 149:253–260

Fernandez A, Torres-Giner JML (2009) Novel route to stabilization of bioactive antioxidants by encapsulation in electrospun fibers of zein prolamine. Food Hydrocoll 23:1427–14332

Gao WL, Li N, Zong MH (2013) Enzymatic regioselective acylation of nucleosides in biomass-derived 2-methyltetrahydrofuran: kinetic study and enzyme substrate recognition. Bioresour Technol 164:91–96

Gouveia L, Oliveira AC (2009) Microalgae as a raw material for biofuels production. J Ind Microbiol Biotechnol 6:269–274

Guha R, Vinjamur M, Mukhopadhyay M (2011) Demonstration of mechanisms for coprecipitation and encapsulation by supercritical antisolvent process. Ind Eng Res 50:1079–1088

Halim R, Gladman B, Danquah MK, Webley PA (2011) Oil extraction from microalgae for biodiesel production. Bioresour Technol 102:178–185

Ibanez MEE, Cifuentes A, Rodriguez-Meizoso I, Mendiola JA, Reglero G, Senorans J, Turner C (2009) Spanish Patent no. P200900164

Jafari SM, Assadpoor E, He Y, Bhandari B (2008) Encapsulation efficiency of food flavours and oils during spray drying. Drying Technol 26:816–835

King JW, Srinivas K (2009) Multiple unit processing using sub-and supercritical fluids. J Supercrit Fluids 47:598–610

Lohrey C, Kochergin V (2012) Biodiesel production from microalgae: co-location with sugar mills. Bioresour Technol 108:76–82

Mezzomo N, Paz E, Maraschin M, Martín A, Cocero MJ, Ferreira SRS (2012) Supercritical anti-solvent precipitation of carotenoid fraction from pink shrimp residue: effect of operational conditions on encapsulation efficiency. J Supercrit Fluids 66:342–349

Palavra AMF, Coelho JP, Barroso JG, Rauter AP, Fareleira JMNA, Mainar A, Urieta JS, Nobre BP, Gouveia L, Mendes RL, Cabral JMS, Novais JM (2011) Supercritical carbon dioxide extraction of bioactive compounds from microalgae and volatile oils from aromatic plants. J Supercrit Fluids 60:21–27

Pereira CG, Meireles MAA (2010) Supercritical fluid extraction of bioactive compounds: fundamentals, applications and economic perspectives. Food Bioprocess Tech 3:340–372

Pignolet O, Jubeau S, Vaca-Garcia C, Michaud P (2013) Highly valuable microalgae: biochemical and topological aspects. J Ind Microbiol Biotechnol 40:781–796

Prado JM, Prado GHC, Meireles MAA (2011) Scale-up study of supercritical fluid extraction process for clove and sugarcane residue. J Supercrit Fluids 56:231–237

Rostagno MA, D'Arrigo M, Martínez JA (2010) Combinatory and hyphenated sample preparation for the determination of bioactive compounds in foods. Trends Anal Chem 29:553

Ruen-ngam D, Shotipruk A, Pavasant P, Machmudah S, Goto M (2012) Selective extraction of lutein from alcohol treated *Chlorella vulgaris* by supercritical CO_2. Chem Eng Technol 25:255–260

Santos DT, Meireles MAA (2010) Carotenoid pigments encapsulation: fundamentals, techniques and recent trends. Open Chem Eng J 4:42–50

Santos DT, Meireles MAA (2013) Micronization and encapsulation of functional pigments using supercritical carbon dioxide. J Food Process Eng 36:36–49

Santos DT, Albuquerque CLC, Meireles MAA (2011) Antioxidant dye and pigment extraction using a homemade pressurized solvent extraction system. Procedia Food Sci 1:1581–1588

Santos DT, Barbosa DF, Broccolo K, Gomes MTMS, Vardanega R, Meireles MAA (2012a) Pressurized organic solvent extraction with on-line particle formation by supercritical anti solvent processes. Food Public Health 2:231–240

Santos DT, Veggi PC, Meireles MAA (2012b) Optimization and economic evaluation of pressurized liquid extraction of phenolic compounds from jabuticaba skins. J Food Eng 108:444–452

Semeonova SI, Ohya H, Higashijima T, Negishi Y (1992) Separation of supercritical CO_2 and ethanol mixtures with an asymmetric polyimide membrane. J Membr Sci 74:131–139

Sosa MV, Rodrıguez-Rojo S, Mattea F, Cismondi M, Cocero MJ (2011) Green tea encapsulation by means of high pressure antisolvent coprecipitation. J Supercrit Fluids 56:304–311

Temelli F (2009) Perspectives on supercritical fluid processing of fats and oils. J Supercrit Fluids 47:583–590

Vanthoor-Koopmans M, Wijffels RH, Barbosa MJ, Eppink MHM (2013) Biorefinery of microalgae for food and fuel. Bioresour Technol 135:142–149

Vatai T, Skerget M, Knez Z (2009) Extraction of phenolic compounds from elder berry and different grape marc varieties using organic solvents and/or supercritical carbon dioxide. J Food Eng 90:246–254

Yen HW, Hu IC, Chen CY, Ho SH, Lee DJ, Chang JS (2013) Microalgae-based biorefinery–from biofuels to natural products. Bioresour Technol 135:166–174

Yesil-Celiktas O, Cetin-Uyanikgil EO (2012) In vitro release kinetics of polycaprolactone encapsulated plant extract fabricated by supercritical antisolvent process and solvent evaporation method. J Supercrit Fluids 62:219–225

Yoo G, Park WK, Kim CW, Choi YE, Yang JW (2012) Direct lipid extraction from wet *Chlamydomonas reinhardtii* biomass using osmotic shock. Bioresour Technol 123:717–722

Zougagh M, Valcarcel M, Rıos A (2004) Supercritical fluid extraction: a critical review of its analytical usefulness. TrAC Trends Anal Che 23:339–405

Printed in the United States
By Bookmasters